THE DEMONSTRATION SOCIETY

Infrastructures Series

edited by Geoffrey C. Bowker and Paul N. Edwards

A list of books in the series appears at the back of the book.

THE DEMONSTRATION SOCIETY

CLAUDE ROSENTAL

TRANSLATED BY CATHERINE PORTER

THE MIT PRESS   CAMBRIDGE, MASSACHUSETTS   LONDON, ENGLAND

© 2021 Massachusetts Institute of Technology

All rights reserved. No part of this book may be reproduced in any form by any electronic or mechanical means (including photocopying, recording, or information storage and retrieval) without permission in writing from the publisher.

The MIT Press would like to thank the anonymous peer reviewers who provided comments on drafts of this book. The generous work of academic experts is essential for establishing the authority and quality of our publications. We acknowledge with gratitude the contributions of these otherwise uncredited readers.

This publication was supported by:
French National Center for Scientific Research (CNRS)

Laboratory of Excellence: Transformation of the State, Politicization of Societies, Institution of the Social (LabEx TEPSIS)

Center for the Study of Social Movements (CEMS)

School of Advanced Studies in the Social Sciences (EHESS)

Treilles Foundation

This book was set in ITC Stone and Avenir by New Best-set Typesetters Ltd. Printed and bound in the United States of America.

Library of Congress Cataloging-in-Publication Data

Names: Rosental, Claude, author. | Porter, Catherine, 1941- translator.
Title: The demonstration society / Claude Rosental ; translated by Catherine Porter.
Description: Cambridge : The MIT Press, 2021. | Series: Infrastructures | Includes bibliographical references and index.
Identifiers: LCCN 2020045142 | ISBN 9780262542890 (paperback)
Subjects: LCSH: Demonstrations—Social aspects. | Social movements. | Product demonstrations. | Business presentations.
Classification: LCC HM866 .R6713 2021 | DDC 303.48/4—dc23
LC record available at https://lccn.loc.gov/2020045142

10 9 8 7 6 5 4 3 2 1

They are whole "entities," entire social systems, the functioning of which we have attempted to describe. . . . It is by considering the whole entity that we could perceive what is essential, the way everything moves, the living aspect, the fleeting moment when society, or men, become sentimentally aware of themselves and of their situation in relation to others. In this concrete observation of social life lies the means of discovering new facts, which we are only beginning dimly to perceive.

—Marcel Mauss, *The Gift* ([1925] 1990), 100

CONTENTS

PREFACE ix
ACKNOWLEDGMENTS xiii

INTRODUCTION 1

1 A DRAMATURGIC APPROACH TO PUBLIC DEMONSTRATIONS 19
2 PUBLIC DEMONSTRATIONS IN ACTION 37
3 PREPARING DEMONSTRATIONS 79
4 CAMPAIGNS AND ARSENALS 105
5 PERFORMANCES WITH LIMITED POWER 133
6 APPRENTICESHIP AND PROFESSIONALISM 167
7 ENTERTAINMENT, CULTURAL GOODS, FORMS OF ART 193

CONCLUSION 227

NOTES 237
REFERENCES 249
INDEX 263

PREFACE

Demos of makeup products on YouTube by famous influencers, public demonstrations by nongovernmental organizations to prove the existence of environmental dangers, demonstrations of military might in North Korea: public demonstrations are omnipresent in contemporary societies. Instruments of proof and persuasion among others, these artifacts and performances may be found not only in educational establishments, research institutes, and courtrooms, but also in the streets or in sites of commercial exchange. Their preparation and execution can mobilize many resources, create considerable tension, and arouse strong passions. They constitute intense moments of social life that involve on a broad scale alliances, material and symbolic possessions, and more generally the future of individuals and of groups.

While their social, political, and economic stakes are considerable, demonstrations are often perceived as isolated, even anecdotal events, and therefore hardly worthy of systematic analysis. Reversing the perspective, this book puts public demonstrations at the heart of the analysis. It examines the various practices associated with the concept, their common features, their particularities, and their sometimes complementary uses. It attempts to show to what extent it is productive to examine public demonstrations as a whole. The book offers detailed analyses of the processes through which demonstrations are conceived and deployed in public, and of the many skills of their creators, relying on the results of a large number of investigations carried out in various sociohistorical spaces (e.g., at information technology conferences, on the internet, in business offices, at research institutions).

Some of these results have been obtained by researchers working in various domains of the social sciences, especially historians. But most stem from investigations I carried out during the past thirty years, during which I observed the uses of public demonstrations throughout the world: high-tech demos in Silicon Valley, major street protests in France, demonstrations of research programs launched by the European Commission in Brussels, demonstrations of faith during religious processions in Córdoba, and even performances by skilled demonstrators at the Paris Fair. Over the years, I have also compared job listings for demonstrator-salesmen on both sides of the Atlantic, have analyzed the advice of software evangelists addressed to would-be "demo gods," and have studied the role allocated to "demonstrative activities" in the project of the European constitution.

On the basis of these rich materials, I show in particular how public demonstrations are produced by central authorities, interest groups, organizations, and a multitude of individuals in efforts to lead the world, develop and sell products, and build markets. I also show how public demonstrations represent fundamental forms of interaction and exchange that can consist, for instance, in ceremonial events, commercial spectacles, or artistic enterprises.

By putting together the pieces of a vast puzzle, I attempt to bring to light the pervasiveness of public demonstrations in societies, both in the contemporary period and throughout history. Individuals in today's societies, like those of older societies, are or have been confronted in their daily lives with a multitude of public demonstrations in diverse forms, when they are not, or have not been, the authors of such events themselves.

The scope of these phenomena suggest that we are all living today, as in the past, in a society of demonstration. This society is marked by the existence of an imperative to demonstrate that is deployed on multiple scales and on many different bases. Taken as a whole, the phenomena in question are at once economic, political, and aesthetic in nature. They are of the cognitive, legal, and religious order. They involve domestic life and professional life. They implicate all social classes.

Even as it illustrates the ways in which public demonstrations structure social, economic, and anthropological exchanges, the book attempts to make it clear how these demonstrations represent total social facts that encompass all institutions. It also suggests why these practices are at the origin of regimes that can be called demo-cratic, to the extent that they mobilize demonstrations as a way of managing public affairs.

In dialogue with many other works in the social sciences, ranging from the sociology of social movements to economic anthropology, from political philosophy to the history of science and technology, this book is designed to offer a new guiding thread that might lead toward a better grasp of fundamental but as yet largely unseen dynamics of contemporary societies.

ACKNOWLEDGMENTS

I thank the CNRS (the French National Center for Scientific Research) for allowing me to pursue the research that led to the writing of this book. I am also grateful for support for the translation from several other French organizations: the LabEx TEPSIS (Laboratory of Excellence: Transformation of the State, Politicization of Societies, Institution of the Social), the CEMS (Center for the Study of Social Movements), and the EHESS (School of Advanced Studies in the Social Sciences). I also benefited from a period of residency devoted to writing at the Fondation des Treilles (Treilles Foundation).

I also wish to express my gratitude to all the colleagues and friends from whose remarks and constructive criticism I have benefited in the course of various seminars, colloquia, and informal discussions, and who in many different ways have helped me bring this project to fruition. While I cannot name them all, I should like to address special thanks to Charlotte Bigg, Luc Boltanski, Sabine Chalvon-Demersay, Karine Chemla, Eve Chiapello, Monique Dagnaud, Nicolas Dodier, Yves Gingras, Neil Gross, Sheila Jasanoff, Laurent Jeanpierre, Laurence Kaufman, Bernard Lahire, Michèle Lamont, Jérôme Lamy, Michael Loriaux, Grégoire Mallard, Nortje Marres, Chandra Mukerji, Louis Quéré, Arnaud Saint-Martin,

Steven Shapin, Philippe Sormani, David Stark, Marie Thébaud-Sorger, Sezin Topçu, and Julia Velkovska. I also would like to thank Geoffrey Bowker, Paul Edwards, Justin Kehoe, and the whole editorial team at MIT Press for their editorial support and advice, as well as Catherine Porter for her splendid translation. Finally, I thank my family and friends for all the forms of support they have given me, and I dedicate this work to the memory of my parents.

INTRODUCTION

Steve Jobs's highly publicized presentations of new Apple products, tutorials in the use of cosmetics on YouTube, demonstrations showing how featured household appliances work in stores and marketplaces or on television shopping networks and the internet: "demos" are omnipresent in contemporary societies. So are many other forms of public demonstrations. If businesses are theaters for innumerable PowerPoint presentations and displays of skill, the media relay a great many public demonstrations produced by scientists, lawyers, or politicians—for instance, US Secretary of State Colin Powell demonstrating the existence of weapons of mass destruction in Iraq in 2003 at the United Nations headquarters. Urban spaces do their part, harboring countless demonstrations of athletic and artistic prowess, along with street protests of all sorts. According to a recent estimate, on average three street protests a day are staged in Paris alone (Fillieule 1999, 199).

Many people are thus confronted on a daily basis by a stream of public demonstrations, and sometimes even enlisted as participants. Social relations, exchanges, and everyday lives in contemporary societies appear to be structured to a significant extent by these practices. While they are often perceived as incidental, since each one is grasped in isolation, their stakes are high.

The project of analyzing them as a coherent whole, and exploring what I came to think of as a "demonstration society," took shape during a long period of preparation. The starting point was my inquiry into the work of logicians and researchers in the United States in the early 1990s. I quickly became intrigued by the time and energy these scientists devoted to preparing and carrying out demonstrations of software and robots, above and beyond their commitment to public presentations of their theoretical results in talks or articles. For example, the scholars working in MIT's Media Lab—a research center famous for its work in the high-tech field—produced multiple public demonstrations of technology ("demos") for their peers, or for sponsors invited to tour the laboratory. Nicolas Negroponte, the center's founder, had even made the slogan "demo or die" a watchword within the lab. The formula of course played on the famous "publish or perish" dictum in the academic world, which stresses the imperative of publication in the pursuit of a career in university teaching. It was a telling illustration of the essential role of these demonstrative practices in the life of that particular research laboratory (Brand 1987).

In Silicon Valley, the hub of new technologies located near San Francisco, it was ordinary practice for researchers and engineers in information technology to introduce themselves by offering to "do a demo" of their accomplishments. I observed frequent recourse to demos within various organizations in the region, especially at Stanford University, as well as at major conferences on artificial intelligence throughout the United States (Rosental 2007, 2008). Similarly, in the late 1990s, during a study of management practices at a major research and development program of the European Commission in the realm of information technologies, I was struck by the importance of the use of demos by engineers and managers in telecommunication and IT companies before audiences consisting of people with political and economic responsibilities (Rosental 2005, 2017).

These phenomena brought a number of questions to mind. In particular, what stakes might explain why such varied actors invest in these apparently playful practices? What sociohistorical conditions could explain the development of such intensive use of performances that were often perceived as anodyne or even insignificant? And what were the connections and the differences between these performances and other public demonstrations that took place in different forms in multiple social spaces, such as, for example, the demonstrations of kitchen tools made by voluble demonstrators in marketplaces or on television screens?

Over the years, these questions led me to analyze the ins and outs of demos in my own fields of inquiry. I gradually extended my investigation to a wide range of demonstrative practices, ranging from the organization and unfolding of major street protests in France, such as the *"Sciences en marche"* (Scientists on the March) event that took place in Paris in 2014,[1] to demonstrations by exhibitors at the Paris Fair[2] or at showcases for entrepreneurs (e.g., Microsoft's TechDays), and even protests by theater professionals in the streets of Avignon during the annual festival.[3] I also looked into various athletic and artistic achievements accomplished in urban spaces, demonstrations of faith in religious processions, and demonstrations designed to promote sales in commercial spaces such as information technology stores.

Ethnographic observations and interviews have been supplemented by research on the internet, now a privileged site for disseminating public demonstrations (especially in video format), and on various sets of documents. Thus, for example, I have compared the contents of job listings for sales representative-demonstrators on both sides of the Atlantic, and I have analyzed advice offered to "technology evangelists"[4] on how to become a "Demo God."

I then compared the results of these various investigations to those produced by social scientists in works scattered throughout a vast literature, both contemporary and historical. Those I have identified help document the practices of public demonstrations, even if such demonstrations were not always central to the writers' concerns. In the course of my research, the potential impact and the multiple stakes of public demonstrations in societies became increasingly clear, as did the usefulness of an initial synthesis.

WHAT IS A PUBLIC DEMONSTRATION?

To begin with, I need to indicate what is meant here by the term "public demonstration." While the foregoing examples sketch out a space of signification for the expression, it is time to specify its contours.

I should note first of all that, in the scientific realm, the meanings attributed to the term have gone through important evolutions and variations as the sociohistorical context has evolved.⁵ Among these contexts we can include, for example, Galen's anatomy demonstrations in antiquity, the public dissections of cadavers and the elaborate drawings in books known as "theaters of machines" in the Renaissance, and seventeenth-century demonstrations of natural philosophy such as Boyle's public experiments with air pumps (Dolza and Vérin 2003; Shapin and Schaffer 1985; Von Staden 1994; Van Dijck 2001). We can also include public lectures on Newton's laws in the eighteenth century, which used various instruments such as Atwood's machine, and practices involving magic lanterns (an early form of slide projector) in the nineteenth century (Hankins and Silverman 1995; Schaffer 1994).

Existing analyses of the nature and roles of such demonstrations cannot be reduced to the syllogistic Aristotelian conceptions that largely predominated up to and including medieval philosophy (Lloyd 1990). Depending on the context, scholars have accounted for demonstrations by invoking notions of proof from the realm of geometry, or by characterizing them as tools of persuasion, as pedagogical instruments, as a form of jousting, or even as entertainment. This multiplicity and variability of characterization has continued into the contemporary period. In order to present a synthetic viewpoint, while extending the focus beyond the sciences, it will be useful to analyze each element of the expression "public demonstration" separately.

"Demonstration," from the Latin *demonstratio*, has several meanings according to the context in which it is used. In French, the word is commonly defined in three ways: (1) as a proof (or an undertaking aiming to establish the truth of a proposition), especially in the sciences; (2) as an action consisting in showing something in order to explain, interest, or impress (showing how a device works, displaying an athletic, artistic, or technological skill, shifting the position of armed forces in order to intimidate or deceive an adversary); or (3) as the manifestation of an inclination, an intention, or an emotion. In English, the term has the same Latin origin and meanings as in French, but since the mid-nineteenth century it has taken on a supplementary meaning, as (4) a public gathering, rally, or march intended to protest or express an opinion on a political issue.⁶

It thus refers to what is called a *manifestation* in French (colloquially, a *manif*), that is, a gathering organized in a specific place aiming to defend a viewpoint or to express a demand.

The phenomena corresponding to this fourth definition have been characterized in a variety of ways in the sociological literature dealing with social movements (Callon 2003; Cefaï 2007; Favre 1990; Fillieule 1999; Tarrow 1994; Tilly 1986; Traïni 2010). Depending on the analyst, the notion of demonstration in this fourth sense refers to more or less circumscribed courses of action such as collective marches, occupations or obstructions of public or private spaces, or protest meetings, violent or not. It can also refer to strikes, petitions, and various noninstitutional public actions. The criteria for identifying this sort of demonstration include not only the type of action undertaken but also its more or less "political" character, the nature and number of participants, organizers, and observers, as well as the type of message conveyed—for example, the expression of an opinion or a demand.

The word "demonstration" thus refers to a set of objects (the text of a mathematical proof, for example) and practices (such as a demonstration of an ethnic dance) that are organized a priori, according to their customary meanings, in the four subgroups previously identified, categories whose nature, modes of articulation, and differences warrant detailed exploration. The term thus appears to be no more univocal than many others; the notions of "institution," "people," "farmer," "social movement," or "finance," for example, all refer to a multitude of heterogeneous objects and practices. They give rise to more or less exclusive, imprecise, and competing definitions (Champagne 1984; Tournay 2014). Moreover, the term "demonstration" is not endowed with a space of signification based on a strong institutional structuring, as is the case, for example, with the notion of "farmer." The latter benefits, in various countries, from the existence of professional organizations and unions, and from an associated statistical category. Still, the term "demonstration" does not refer to everything or just anything at all. Like other notions, it lends itself to analysis as a specific set of phenomena and objects.

Instead of trying at the outset to reduce the four subdefinitions I have listed to a single formulation or to deconstruct them in a radical way, I shall use them to launch the first stage in my analysis. It would be out of the question to try to establish once and for all an overview that would cover all possibilities; I propose, rather, to set down initial reference points in the process of delimiting a field of study. A given object or performance can be described as a demonstration, or not, depending on the actors who perceive it, and according to the historical moments or periods involved. For example, a plea made at a jury trial may be understood by some listeners as a demonstration that a crime has been committed, and by others as a discourse offering no proof on the matter. Moreover, individual points of view on such a matter may well evolve over time (Dulong 1998). This initial effort to define the term "demonstration" thus needs to be extended by consideration of the contradictory undertakings of interpretation and qualification brought to bear by the actors involved.

But we need to consider the term "public," as well, and consider what that epithet can mean, as applied to a "demonstration." A demonstration can be called public to the extent that it is carried out in the presence of several individuals or conceived to transmit a message to a particular group (Goffman 1963). The notion of public may also refer, in the context of demonstrations, to the institution of a common space, to an act of putting on display, or to access to a broad or even unlimited audience. In such cases, steps taken to reach "a public" can be contrasted with steps that might be taken to protect a secret or to maintain privacy; they can also be contrasted with an absence of communication, or with various forms of appropriation (Fraser 1992; Quéré 1992, 83–89). The notion of public can also apply to the possibility for individuals to conceive of themselves as part of a collective in the eyes of a third party, as in the case of individuals who come together for a political demonstration.

These various dimensions are not mutually exclusive. If we consider them all—we are concerned here with demonstrations that can be considered public in one or another of these senses—the term "public demonstration" appears to refer to a circumscribed set of practices and objects of quite distinct natures, which can be brought together a priori in several subgroups, as we have seen with the term "demonstration."

"Public demonstration" characterizes, in the first place, proofs made public (for example, mathematical proofs published in journals) or deployed in public, whether, for example, by lawyers before juries, by scientists before gatherings of their peers, or by politicians before the media. This expression also refers to several types of performances before various audiences, most notably in the context of sports (for instance, a demonstration of martial arts) or the performing arts (for instance, a demonstration of hip-hop). It also refers to presentations designed to teach (demonstrations of milking in dairy farms open to the public), demonstrations of strength (military parades or maneuvers intended to impress or deceive an adversary—see figure 0.1) and various sorts of demonstrations before publics of diverse shapes and sizes: manifestations of faith, inclinations, intentions, or emotion, but also—and more commonly in English and in German than in French—demonstrations in the sense of public gatherings intended to express a demand or political opinions.

Finally, the term can correspond to presentations of various technological products (software, household appliances, and so on). Also sometimes called "demos," performances of this type can serve pedagogical objectives, foster sales, or simply exhibit the competency of the demonstrator.

0.1 Demonstration of military might in North Korea in 2018. © 2018 Tasnim News Agency. Creative Commons Attribution 4.0 International license (https://creativecommons.org/licenses/by/4.0/deed.en).

Demos of this sort can be carried out in marketplaces, stores, or within businesses. They can also take place in exhibit halls at high-tech conferences, in a domestic framework, or on telemarketing platforms. In this last case, as in many others, the goal is often not the realization of a live performance but rather the transmission of a video recording, which can occur on a very large scale, either on television or on the internet.

The authors of these performances may be trade representatives, market demonstrators, consultants, start-up creators, scientists, or professional actors, among other possibilities. Demos can also be produced by machines programmed to explain how they work automatically. Software for video games in particular often has a "demo mode" that shows users or potential buyers how it works without requiring the presence of a human operator.

In these diverse cases, the notion of "presentation" is evocative. The point is to bring something into the presence of a public: an exceptional phenomenon, a device in action, or a capability, something that is presumed to be not obvious at first glance, something abstract if not intangible. Depending on how broadly one defines technology, the word "demo" can be used to characterize a larger or smaller number of public demonstrations, most notably in the sciences. At the same time, a demonstration using technology is not necessarily a demo. A public PowerPoint software presentation does not generally constitute a demonstration of the way the software works and thus does not count as a demo, although a demo may well have been carried out when PowerPoint was introduced.

Public demonstrations thus take place in many different spaces. As I have suggested, these performances and the objects on display proliferate first of all in the realm of economic exchanges. Demos serve in particular to launch new technologies (Simakova 2010), as in the case of the highly publicized presentation of Apple products by Steve Jobs (see figure 0.2). They are used to sell products in various spaces—markets, stores, home settings with Tupperware parties and the like (Clark and Pinch 1992, 1995; Le Velly 2007; Pinch and Trocco 2002; Pinch 2003; Sherry 1998) or to support projects that involve a variety of technologies (Capelle 2012; Rosental 2002; Callon and Muniesa 2007). On this basis, they are found, for example, in the health field (Coopmans 2011; Winthereik, Johannsen, and Strand 2008), and in architecture (Houdart 2005; Yaneva 2009).

0.2 Steve Jobs demonstrating the MacBook Air at the Macworld Conference in 2008. © 2008 Aljawad. Creative Commons Attribution-Share Alike 3.0 Unported license (https://creativecommons.org/licenses/by-sa/3.0/deed.en).

In the sciences, public demonstrations are used most notably for the purposes of proof, illustration, or teaching (Rosental 2004, 2007). In information technology or in the sector of video games, they bring together communities of users (Auray 1997). In politics, they are used in particular to persuade or to test public support, to mobilize groups of various sizes, or to legitimize government actions (Barry 2001; Brian 2001; Girard and Stark 2007; Mukerji 1997; Rosental 2011). They are found in abundance in religious practices and texts, chiefly as proofs and manifestations of the divine. In the realm of cultural production, they are used among other things to entertain crowds (Vidal 2006; Lunenfeld 2016).

Given the diversity of the contexts in which public demonstrations occur, it is not astonishing that studies focused on them are themselves dispersed among several fields within the social sciences. While the sociology of social movements has concentrated primarily on analyzing the modes of political protest, participation, and intervention that demonstrations convey, economic sociology has looked chiefly at their role in sales and marketing practices. Whereas interactionist sociology has paid particular attention to the dramatic dimension of such performances, the uses of public demonstrations in research and teaching, as well as their epistemological status, have been examined above all by specialists in the sociology, philosophy, and history of science and technology.

This splintering of the problematics has not tended to encourage an overall examination of the results of studies of public demonstrations, and still less the development of a research field devoted to the subject. Given the diversity of objects and practices associated with demonstrations, one may well wonder why it would be appropriate to analyze them together. There are several possible reasons to consider.

WHY STUDY PUBLIC DEMONSTRATIONS AS A GROUP?

In the first place, a study devoted to all forms of public demonstration makes it possible to look into their common features and their differences in the various spaces where they occur. For example, we can ask what unifies or differentiates the practices of demos in a marketplace and those of street protests, in terms of the way they are staged, for example, or the way the demonstrators manage interactions with the public, or the way interpretations of the performances are controlled. We can ask to what extent the techniques that organizers of political demonstrations use to influence the contents of journalists' articles are comparable to the techniques that demonstrators of household products use to influence the judgments of potential buyers. By adopting a comparative approach, we may be able to produce a structured set of results derived from a series of very diverse practices and objects, whose most visible connection comes from the use of a common label by actors who may operate in widely separated spaces.

The recourse, by certain actors, to several different forms of public demonstration, either successively or simultaneously, also argues in favor of a global study of these forms. For example, NGOs (nongovernmental organizations) such as Greenpeace regularly organize both occupations of particular sites and public demonstrations designed to prove the existence of environmental dangers. The latter often take the form of press conferences focused on "scientific facts," such as the publication of radioactivity levels measured at the site of nuclear waste depositories (Lemieux

2008). Similarly, in order to demonstrate and publicize their results, many researchers in artificial intelligence rely on software demos as much as on written documents published in academic journals or shared in oral presentations in the context of seminars (Rosental 2007, 2009).

To the combined uses of various forms of public demonstration, owing to their complementary roles or their variable degree of relevance according to the situation, we can add the effects of substituting one form of demonstration for another in different spaces. For example, in recent years, recourse to computer-based demos has partially replaced oral demonstrations during thesis defenses in the field of logic (Rosental 2008).

The debates and confrontations that lead actors to call on different forms of public demonstration to defend their respective viewpoints also justify an overall study of these forms. For example, from the postwar period on, certain opponents of nuclear power have used public protests and arguments relayed by the media to challenge institutions seeking to demonstrate the harmlessness and the benefits of this type of energy; institutional actions have often involved organizing visits to power plants for the general public and offering various types of demonstrations produced by engineers dealing with nuclear technologies (Topçu 2013).

The circulation of actors in a variety of social spaces is moreover a source of migration, adaptation, or mutation of demonstrative practices. For example, certain actors in the field of artificial intelligence in Silicon Valley have over time or simultaneously worn the hats of university professor, entrepreneur, consultant, high-tech company employee, defense department advisor, or trade representative with the status of technical commercial engineer. These shifting contexts have led actors to rely on demos in the various areas of their professional lives, whether the objective was to present technological achievements, sell high-tech products, or illustrate theoretical results. The actors' use of demos could be inspired by the work they had done to produce proofs of theorems for oral presentations and publications in an academic context, but it could also be influenced for example by their reading of works in the field of marketing that explained in detail how to create demos (Rosental 2007). Here again, if we are to be able to grasp the associated phenomena of hybridization, an overall attention to the various forms of public demonstration appears essential.

Such attention is also crucial for apprehending the sometimes multiple roles played by public demonstrations. For example, a given demo simultaneously may be envisaged as a public proof, a means for mobilizing crowds, and an event to promote the sale of a product. There is no reason to suppose that a given demo plays only a single role; each demo should be construed rather as a potentially multifunctional object. By the same token, if we are studying collective mobilizations, for example, it may be helpful not to focus a priori on street protests to the exclusion of other forms of demonstration, such as demos. A similar argument can be made for the study of presentations of public proofs or of promotional undertakings: on these occasions, too, various forms of demonstration can be called upon.

The decision not to focus our analysis on certain forms of public demonstration to the exclusion of others makes it possible, moreover, to question the constraints that weigh on the exercise of all forms, and on their conditions of possibility. From this standpoint, we can try to understand, for example, why representatives of an organization such as Greenpeace choose, in a given context, to occupy a site rather than to demonstrate a scientific fact before a gathering of journalists, if we analyze the potential maneuvering room available to the organizers for making such a choice, and the elements most apt to determine that choice.

A study bearing on all forms of public demonstration thus appears useful from a variety of standpoints. Admittedly, the interest of such a study is not immediately obvious; indeed, it only became clear to me over time, after I had undertaken a set of investigations into several forms of demonstration in different spaces. For, in addition to the reasons I have just invoked, I must add an entirely different series of motivating factors, which became apparent to me for the most part after the fact, once I had begun my investigation.

For one thing, the diversity and potency of the stakes in public demonstrations in social life make their objects and practices a prime instrument for x-raying contemporary and past societies. Since public demonstrations are sometimes produced and used on an industrial scale, their study helps reveal a number of important social dynamics.

A systematic study of the various forms of public demonstration thus allows us to connect the results obtained from multiple fields of inquiry, beginning with the problematics that are inscribed in various areas of the social sciences, and in sociology in particular (the sociology of organizations, the sociology of science and technology, the sociology of politics, of culture, and so on). In the field of sociology today, such a study offers a rare opportunity to go beyond the framework of specialization to reconnect with certain ambitions that marked the beginnings of the discipline. It also makes it possible to transform the vantage point and bring to light processes that are hard to see when they are encountered in isolation, but that are easier to spot and more meaningful when they appear less singular.

These observations imply, for specialists and general readers alike, the need to bracket our a priori representations of public demonstrations and of what might be at their core: proofs, tools for persuasion, instruments for mobilization, pedagogical support structures, theatrical performances, and so on. We shall need, on the contrary, to be open to being surprised by the diversity of their contents and their roles.

It may be useful, at this stage, to specify that such an enterprise does not define a functionalist horizon as such. It is in fact possible to seek to identify the roles played by public demonstrations for different individuals and institutions without necessarily raising questions about the potential functions these demonstrations might have for society taken as an organism; such questions are not in the purview of the present study.

Moreover, as I have already suggested, I do not mean to assert that we are dealing with a field of homogenous practices and objects, that we should for example consider the presentation of a proof by a mathematician at the blackboard in front of his peers in a research seminar as being of the same nature as a demonstration in the street for the defense of some cause. Embarking on a global study does not mean mixing everything together and lapsing inevitably into reductionism. Nor do I mean to assert that we are facing a series of practices and objects of such diversity that anything or almost anything can be viewed as a public demonstration. By studying the more or less clear and consensual demarcations produced by the actors themselves around the notion of public demonstrations, I am on the contrary seeking to circumscribe a field of practices and objects structured in such a way as to be susceptible to sociological analysis.

DATA AND METHODS

As I indicated at the outset, my reflections will be based on the results of investigations I have undertaken since the early 1990s in Europe and in the United States, along with the results of complementary studies carried out by others throughout the world. My analysis will also rely on several case studies that have punctuated the history of science and technology for a very long time; reckoning with these studies will allow me to bring to light both convergences and divergences in approach, and to spell out the scope and limits of my own inquiry. I am clearly not about to propose an exhaustive analysis of a field of practices and objects whose breadth exceeds the possibilities of even a team of investigators, no matter how large. I propose, rather, to set forth a considerable and coherent body of data in order to produce an initial systematic analysis of the phenomena I have taken as my object, and to document in particular the modes of preparation and achievement that underlie various types of demonstrative performances.

As I have suggested, the constitution of this corpus and the choices that the process entailed did not stem from a research program planned at the outset, but rather from a highly iterative undertaking that has called at the end of its run for a synthesizing exercise. My inquiries and the exploration of various fields of literature have raised questions along the way that have led me to carry out new investigations and new bibliographical research. The precise contours that shape this work took many years to define; while I shall not attempt to retrace the logic and the exact path I have followed, I should specify that once those contours appeared with sufficient clarity, the choices I made derived from a desire to analyze significant cases of public demonstration, taking into account in particular the definitions associated with the term. Studying the available work on the topic led me either to carry out field work in areas that had thus far been little studied, if at all, or on the contrary to undertake new investigations into the most widely studied cases (for example, fairground demonstrations) to corroborate, nuance, or supplement the available results.

Among the data on which I rely, one set deals with the organization and development in recent years of street protests in France, and tangentially in other European countries and in the United States: in particular, demonstrations by scientists, farmers, and professionals in the performing arts, plus others organized by anti-nuclear militants, anti-bullfighting protestors, and objectors to highway development.

A second set of data concerns demos carried out chiefly in the Western world. Some come from sites of sales (markets, fairs, supermarkets), from academic spaces (universities, research centers, academic meetings), or from trade shows. Others come from businesses, banks, architecture firms, or industrial sites such as power plants. Still others come from rural spaces (demonstration farms or sites demonstrating ecological lifestyles, for example), from urban spaces (as in street demonstrations of athletic or artistic prowess), or from private homes. Others, finally, come from television (in the context of telemarketing programs, for instance), from the internet (directed especially toward a public of specialists in information technology, hackers, or gamers), or from the context of contemporary art exhibits.

Another set of results, which overlaps in part with the previous ones, bears on various types of public demonstrations carried out in the field of science and technology, both in the contemporary period and throughout history. They chiefly take the form of demos, public experiments, oral presentations, or published proofs. They concern the deductive sciences as well as the experimental sciences and technologies. Depending on the context, they are intended for peers, students, or nonspecialists interested in the sciences; they may also be addressed to representatives of economic and political power, to the public at large, and to journalists.

The corpus retained also includes various forms of public demonstrations that have been produced in the context of religious processions, PowerPoint presentations, political meetings, or demonstrations of civil or military force. All these data, whether produced by my own research or produced and analyzed by others, have been collected with the help of several methods: principally ethnographic observations, interviews, archival work, and documentary research of various types, the latter including examination of records of meetings, correspondence, and newspaper articles. They also include job listings and audiovisual data disseminated on the internet and through social networks (for example, videos of demos and street protests, PowerPoint presentations, reactions to demonstrations expressed via Twitter). These digital arenas in fact represent important areas of circulation of public demonstrations and formulation of commentaries about these demonstrations. In particular, they make it possible to objectify individual and collective modes of apprehending demonstrations, modes that are much harder to grasp in other spaces.

To avoid burdening this introduction further, I shall wait to spell out the nature of the investigations and the methods used in more detail as I go along. Most of the cases selected will be analyzed in several stages and in relation to one another throughout the book, moreover, so it will be possible to pull together the set of results that emerge from these inquiries.

PUBLIC DEMONSTRATIONS IN PRACTICE

It is easy to imagine that it will take more than one book to analyze the corpus that I have just evoked. This volume thus constitutes a first stage. In order to begin to constitute the field of research I have outlined, I shall focus at the outset primarily on the practice of public demonstrations. In order to be able to circumscribe their nature, their uses, and their stakes, it seems indispensable to begin by observing and analyzing their modalities of use in a methodical way. Moreover, in order to bring to light the reality of the field of research to be undertaken on these social practices, it is important to bring together nonstylized descriptions of the practices in question. How do public demonstrations come about? How are they conceived and realized? These are the questions that will serve as my starting point, with the prospect of shedding initial sociological light on the phenomena in play.

As I have suggested, interactionist approaches to public demonstrations have led to a certain number of descriptions of these practices in terms of performances, theatrical performance in particular. This analytic vocabulary warrants interrogation. Erving Goffman has unquestionably made one of the most significant contributions to the subject to date. I shall begin, then, by studying his proposal to grasp demonstrations as simulations produced for utilitarian purposes, in the context of a dramaturgical approach to social life. I shall try to determine in particular to what extent demos of various types of products proceed on the basis of interplay between fiction and reality, and how this is experienced by the spectators. In particular, I shall try to see whether conventions of theatrical communication are put to work in this context, and whether they aim to suppress the spectators' critical habits of mind. I shall try to find out whether such conventions promote the complicity of audiences and their support for the causes being defended during street protests or for innovative projects whose future is uncertain, when such projects are magnified by spectacular demos. In this context, I shall problematize the nature of the various publics involved, their modes of constitution, and their modes of participation (chapter 1).

The analysis will continue with a detailed examination of the way public demonstrations take place in selected cases. The framework in which demonstrations occur, the way they are carried out, and the exchanges between the demonstrators and their public will be my focus at this point. I shall examine the scenarios adopted, the material arrangements used, and also the gestures deployed, the repertoires invoked, and the ways demonstrators play on emotions (chapter 2).

In addition, I shall explore in detail the preparations, the iterations, and the work of staging involved. I shall analyze the modes through which demonstrators manage their audiences before and after their performances. In particular, I shall study how demonstrators sometimes attempt to elicit expectations among their spectators in advance of the performances and seek to control their interpretations at different stages (chapter 3).

I shall go on to argue that, generally speaking, public demonstrations must not be grasped as isolated acts, but that they are, on the contrary, often inscribed in the framework of demonstrative campaigns that mobilize many forms of demonstration. I shall then analyze the uses of demonstrative arsenals (or counterdemonstrative arsenals, when conflicts develop). Since ex-nihilo creations in this area are far from systematic, I shall study the phenomena of reuse, adjustment, and recombination of demonstrations (chapter 4).

Under these conditions, we shall see that what constitutes the effectiveness and especially the persuasiveness manifested in individual performances often depends on multiple factors that may not be observable in the situation itself. These factors include the prior acquisition of various ploys by the demonstrator, the more or less controlled composition of the audience, and the actions taken before and after a demonstration. Various organizational constraints, incidental effects, or the interplay of competition, or a combination of these, can also play determining roles. Thus, as a general rule, public demonstrations in their various forms seem to be endowed with limited intrinsic power. The reactions they elicit may be variable and sometimes ambivalent. Bugs and cases of "failure," while they are sometimes cleverly turned to the demonstrator's advantage, are legion (chapter 5).

In any event, it is clear that skills in demonstration, and training in those skills, are crucial for ensuring the "success" of a performance. This is just as true for researchers in information technology as for organizers of street protests. We shall see that the tools of the trade can be acquired in many different ways. For the practice of demonstrations is not always an incidental activity, nor does it always reflect just one professional skill among others. It sometimes constitutes a trade of its own, as for example in the case of pharmaceutical sales representatives, or heirs of itinerant peddlers who passed their expertise along (chapter 6).

I shall show that the practice can also be identified in the performing arts and in other forms of entertainment, cultural productions, and artistic undertakings. This is moreover an important reason for acknowledging that a certain number of public demonstrations are in fact performances, theatrical performances in particular, given the extent to which these contexts mark their conception and realization. In short, the practices of public demonstrations are not monolithic. Their diversity has to be grasped in the interstices of a wide range of activities, including ceremonial events, commercial spectacles, or entertaining productions as well as artistic enterprises that seek, for example, to critique marketing methods and the capitalist system (chapter 7).

I propose to begin by looking at the relevance of the dramaturgical approach to public demonstrations developed by Erving Goffman. To this end, it will be useful to examine his proposals in detail before measuring them against the results of the various investigations. This is the object of chapter 1.

1
A DRAMATURGIC APPROACH TO PUBLIC DEMONSTRATIONS

Erving Goffman's work on public demonstrations makes a good starting point for an expanded study of these practices. As I hope to show, Goffman's work invites several major questions about their nature. Can one systematically identify "actors" and "spectators," and to what realities might these terms correspond? What purposes are public demonstrations meant to serve? How do demonstrators and their audiences understand their roles? To what extent are public demonstrations perceived by the public as fiction, as spectacles, as manipulative tools?

In his book *Frame Analysis* (1974), Goffman focused on the interactions that take place during what he calls demonstrations. His analyses bring into play rather technical definitions that may seem somewhat opaque at first glance. I shall begin by attempting to clarify the simple intuitions that they convey.

SIMULATIONS

For Goffman, the term "demonstration" refers to a quite specific set of practices that entail "performance of a task-like activity out of its usual functional context in order to allow someone who is not the performer to obtain a close picture of the doing of the activity" (Goffman 1974, 66). More precisely, the author conceives of demonstrations as a particular type of "technical redoing," that is, "strips of what could have been ordinary activity [that] can be performed, out of their usual context, for utilitarian purposes openly different from those of the original performance, the understanding being that the original outcome of the activity will not occur" (Goffman 1974, 58–59).[1] For Goffman, moreover, these "technical redoings" represent just one type of "keying"; he defines a key as "the set of conventions by which a given activity, one already meaningful in terms of some primary framework, is transformed into something patterned on this activity but seen by the participants to be something quite else. The process of transcription can be called keying" (Goffman 1974, 43–44).

To comprehend these statements, it is helpful to consider Goffman's own illustrative examples, which in practice fall into the category of what are commonly called demos. To support his definitions, Goffman mentions several cases: a vacuum cleaner salesman who shows a housewife how to vacuum up the dirt he has dumped on the floor, a nurse who explains to a mother how to bathe her baby, a demonstrator who shows military personnel how a piece of artillery works, and "a pilot at full altitude [who] shows his passengers what the sound and sensation will be like when air flaps are lowered" (Goffman 1974, 66).

Demonstrations, in Goffman's sense, could thus be qualified as simulations of actions intended to be utilitarian, perceived a priori as such by those who witness them. Unfortunately, the author offers only a hasty analysis of these practices, referring essentially to sequences of action in daily life drawn primarily from newspaper articles. His analysis consists mainly in highlighting social constraints that weigh on these simulations.

Goffman thus indicates that the costs involved in demonstrations should be limited compared to those implied by the simulated action. In some cases, demonstrations should not be overly realistic, especially in cases involving issues of "decency,"[2] for example, in advertisements for deodorants or toilet paper, or in demonstrations of guerrilla tactics and crowd control.

The author's analyses thus supply some initial paths for grasping demonstrations as theatrical performances. In testing their empirical relevance, we shall need to ask whether the costs implied by public demonstrations are always lower than those the activity being simulated would entail.

Questioning Goffman's conceptual frame may also help us conceive of demonstrations as theatrical performances without reducing them to technical redoings. In Goffman's own categorization, technical redoings are just one of several types of keying: he characterizes others as "make-believe," "ceremonials," and "regroundings" (Goffman 1974, 48, 58, 74). It seems, however, that these other types of keying may also characterize some demonstrations.

The mode Goffman calls "make-believe" entails an "activity that participants treat as an avowed, ostensible imitation or running through of less transformed activity, this being done with the knowledge that nothing practical will come of the doing. The 'reason' for engaging in such fantasies is said to come from the immediate satisfaction that the doing offers. A 'pastime' or 'entertainment' is provided" (Goffman 1974, 48). Might not demonstrations of, for example, martial arts for a general audience fall under that category?

A similar question can be raised about "ceremonials," which correspond, for Goffman, to social rituals "such as marriage ceremonies, funerals, and investitures. . . . Like scripted productions, a whole mesh of acts are [sic] plotted in advance, rehearsal of what is to unfold can occur, and an easy distinction can be drawn between rehearsal and 'real' performance" (Goffman 1974, 58). The periodic launchings of new versions of a product such as the iPhone appear a priori as good candidates for such a label. Ceremonials thus constitute an additional resource for conceiving of certain public demonstrations as theatrical performances.

The same holds true for "regroundings," which for Goffman entail "the performance of an activity more or less openly for reasons or motives felt to be radically different from those that govern ordinary actors. The notion of regroundings, then, rests on the assumption that some motives for a deed leave the performer within the normal range of participation, and other motives, especially when stabilized and institutionalized, leave the performer outside the ordinary domain of the activity" (Goffman 1974, 74). To illustrate this category, Goffman evokes the case of "an upper middle class matron" who acts as saleswoman at a church bazaar, and that of "exalted sinners" who take on "lowly tasks performed as penance"

(Goffman 1974, 74–75). As we shall see later on, certain demos carried out by contemporary performance artists fit these descriptions quite well.

Thus, it is clear that there are many reasons to grasp public demonstrations a priori as theatrical performances, and that doing so does not require us to reduce them to simulations intended for utilitarian purposes that would be perceived as such by spectators.[3]

If demonstrations can stem from several different "keys," we can also ask to what extent they may constitute "fabrications" in Goffman's sense, or a mix of keys and fabrications, even if the author does not mention the possibility. For him, a fabrication refers to "the intentional effort of one or more individuals to manage activity so that a party of one or more others will be induced to have a false belief about what it is that is going on" (Goffman 1974, 83). My own studies and those of others (e.g., Smith 2009) suggest that any demonstration can stem not only from one or several of Goffman's keys but also from a fabrication, and that any demonstration may imply a complex relation on the part of the public toward the activities presented. In particular, the public's perception of the art of demonstration may not always be accurate; a demonstrator's efforts to manipulate or mystify may be more or less successful.

For example, during an inquiry I carried out at the Paris Fair in 2014, I was able to observe a demonstration leading to the sale of axes that were supposed to allow people with limited physical abilities to cut substantial pieces of wood with ease. The reactions expressed tended to show that the spectators and buyers had not all realized that the wood used in the demo was quite soft, and that in practice they were likely to be confronted with much more resistant materials.

This example suggests that there is no reason to suppose that the people watching a demonstration always identify precisely what is at stake in the interaction in which they are participating, or that they distinguish clearly and consistently between the model of an activity being presented to them and its reality. This point warrants further development.

FICTION OR REALITY

My own observations and various other studies bring out contrasting pictures of audience ability to distinguish between reality and fiction in the context of demonstrations. We find, first of all, cases such as those analyzed by Goffman, in which the spectators appear conscious both of the simulation being enacted before their eyes and of the utility of the exercise (Rosental 2019). In such cases, one can also speak of "mutually agreed fiction" (Lunenfeld 2000, 13–26).

However, as in the example of the demo at a fair mentioned earlier, members of the audience watching a demonstration may in some cases maintain an ambiguous relation to its fictional character, remaining unaware of it at least in part or forgetting it temporarily or momentarily. This type of phenomenon is well known to scholars who study the practices of spectators of fictional works, in the realm of theater or television series, for example (Chalvon-Demersay 2012). In the latter case, television viewers often identify the actors with the characters they are playing, either while they are watching or in other contexts such as social networks. For this reason, it is generally hard for actors to play a variety of roles on screen.

Comparable phenomena can be observed during scientific demonstrations. In the case of certain public experiments, only specialists manage to perceive the diversity of the artifices at work, whereas at least some members of the general public believe they have direct and unproblematic access to natural phenomena (Collins 1988). Conversely, there are cases in which specialists in one field more or less lose sight of the fictional character of a public demonstration, unlike nonspecialists.

On this topic, I can cite the case of a demo during which the oscillations of an inverted double pendulum were controlled automatically (Rosental 2004). The principle of an inverted pendulum is that it oscillates from a connection located not on its top but rather on its base. It is double when not one but two articulations allow oscillations on two distinct levels.[4] For some researchers in space engineering, this arrangement represents the canonical problem of a rocket at takeoff: it can break apart if the vibrations are not controlled. A control system that makes it possible to limit the oscillations of an inverted double pendulum is thus generally viewed by these specialists as the solution to the problem of launching a spaceship. By contrast, some nonspecialists see it merely as a mechanism of little interest; its role in the launching problem is simply unknown, or else appears artificial. Thomas Kuhn analyzed comparable phenomena in writing about paradigmatic vision, drawing on *Gestalttheorie* (Kuhn 1962; Fleck 1979). The visions in question here cannot be dissociated in any simple way from perceptions of a material reality.

Moreover, certain observations and studies stress the fact that ambiguity, forgetting (even temporary or partial), and the complicity of the public toward the fictional character of a demonstration represent effects that are sought in many demonstrations, and that these effects are permitted by the immersion of the spectators in the play of theatrical conventions. It is a matter of leading the public of a demo, for example, to adopt a less rigorous mode of evaluating a technological device than the one they would adopt in the context of a literal presentation, when they would be seeking complete information and an exhaustive analysis based on multiple criteria.[5] Some demonstrators use this approach to try to get potential clients or investors to have confidence in their promises or their technological utopias in spaces such as showcases for innovation. They seek to deactivate the spectators' critical sense for a time and turn their attention away from the limits or difficulties of a technological project.

Demonstrators can use various procedures to try to obtain this complicity or complacency on the part of the public. They can play on the spectator's emotions and imagination, seeking to arouse the pleasure one feels in going down new paths. Demonstrators can also try to astonish spectators to the point where they will allow themselves to believe in something unexpected. A number of dramaturgical mechanisms and conventions are used to this end by many demonstrators, starting with those who make their presentations on a stage.

Thus, Steve Jobs, for example, rented the San Francisco Symphony Hall in October 1988 to launch the NeXT computer. He hired George Coates, a stage director with experience in multimedia presentations. The décor consisted in a vase of flowers and an unknown object hidden under a black cloth. Steve Jobs made a graphic presentation of the history of information technology and of NeXT to introduce the computer. In the tradition of magicians and daredevil stuntmen, he warned the public that the performance might not work, using that means to highlight the revolutionary character of the machine and to protect himself from the consequences of failure. He spelled it out: "I'd like to remind you of the first two laws of demoing. [Some laughter] First law of demos is that demos will always crash. And the second law of demos is that their probability of crashing goes up with the number of people watching. [More laughter and applause] So if something goes wrong today, have some compassion for the demo-er" (Lampel 2001, 317).

The demo consisted in a presentation of the content of the machine's digital library. It included excerpts from Martin Luther King, Jr.'s "I have a dream" speech and a speech by John F. Kennedy, an audio recording relating to the Apollo 11 moon landing, and selections from musical works. The machine also featured in a duet with a violinist from the San Francisco Symphony, who was present on stage.

This demo did not have a flatly realistic goal. Nor was it simply based on a convention relating to its own fictional character, maintained from start to finish. It aimed at once to foster a relaxed attitude on the part of the spectators, to encourage them to let go of any critical posture, and to make tangible and credible the capacities of the new technology. Simultaneously, it aimed to convince the audience to share in a utopian vision of its possibilities and its uses. Steve Jobs was not the only innovator to have proceeded that way, of course. Thomas Edison, for example, transformed his laboratory in Menlo Park, New Jersey, into a stage from which he could try to persuade multiple actors to electrify America (Bazerman 2002). The launching of the first diesel locomotive, the Pioneer Zephyr, was based in large part on its triumphal arrival, after a record-breaking voyage, at the Chicago World's Fair on May 26, 1934 (Lampel 2001). Similarly, commercial showcases have been used as stage sets for technologies by many innovators lacking the means of a Steve Jobs (Pfaffenberger 1992). This configuration has often presented difficulties for demonstrators, however: it is not always easy to make a product stand out among the other innovations on display, or to achieve the desired level of complicity.

A number of social scientists have stressed the fact that public demonstrations, and demos in particular, can be used to produce illusions that look like realities. Certain studies have focused on recourse to hypnosis, identifying hucksters capable of subjugating their audience (Duval 1981); others have compared crowds of political demonstrators to sleepwalkers under the influence of hypnotists.[6] Still others make it clear that it can be hard to distinguish a demo from an exercise in prestidigitation, in cases where the attention of the spectators is drawn toward certain manipulations in order to make them lose sight of essential realities such as the unsuccessful conclusion of a project or the misfires of a device.

There are abundant examples of public demonstrations of technologies that foreground fictional devices, with the help of tools specially designed for the presentations; some authors speak of "vaporware products" in this connection. The neologism serves to describe phantom software; it was used most notably in the 1990s to describe the software embedded in devices called "personal assistants" that worked correctly only in the context of demos (Markoff 1996).

Two engineers from SunSoft, Annette Wagner and Maria Capucciati (1996), thus attested that they had created a system that had never constituted a real product but that nevertheless functioned as an impressive interface for marketing events. Interviews confirm the recourse, during demos, to software versions containing modules different from the ones destined for the users (Smith 2009). Other case studies focus on demos that do not show the real functionality of a prototype at a given point in time, instead highlighting results that have not yet been achieved or fully confirmed. The status of these demos sometimes appears equivocal, since the audience is not always aware of what has already been achieved and what exactly remains to be done. Such situations are particularly apt to occur when developers are subject to significant pressure to achieve results quickly and to make frequent presentations of their work (Rosental 2007; Coopmans 2011, 721). For an example beyond the realm of demos, we can consider Colin Powell's speech to the United Nations headquarters in 2003, where he tried to demonstrate the presence of weapons of mass destruction in Iraq (Stark and Paravel 2008; see figure 1.1). Similarly, Ronan Le Velly (2007) has described demonstrations at fairs of products presented as if they were made in Europe, in order to emphasize their sturdiness, whereas they were actually made in China. This diversity of configurations makes it possible, moreover, to understand why various sociologists who have studied public demonstrations have described presentations of nonexistent objects or objects that had not (yet) been produced.[7]

We can even find cases of demos conceived explicitly to appear as fictions that were taken as realities by some members of the public. A good

1.1 Colin Powell demonstrating the presence of weapons of mass destruction in Iraq at the United Nations headquarters on February 5, 2003. US Government (https://commons.wikimedia.org/wiki/File:Colin_Powell_anthrax_vial._5_Feb_2003_at_the_UN.jpg).

example is Lauren McCarthy's project, "The Happiness Hat."[8] Created in 2009, the project consisted in an electro-mechanical device coiffed with a bonnet to be worn on a user's head. It was equipped with a receptor that detected whether or not the user was in the process of smiling. When this was not the case, a prod was activated electronically: it caused a painful sensation that ended when the wearer forced a smile.

A demo of this device produced by the artist has been viewed close to half a million times on YouTube and Vimeo, according to Lunenfeld (2016, 349); it has also been broadcast on television. Whereas the demonstration was conceived as an ironic work, it was taken seriously by some portion of the public. Some viewers could not decide whether or not it was meant to be a joke. More than one viewer wrote to the author saying that they were suffering from depression, wanted to order one of the devices, and asked how much it cost.

Descriptions of demonstrations in terms of a dichotomy between fiction and reality thus have to be nuanced, given the variety of cases we may encounter; similarly, it will be useful to interrogate the relevance of the representations of audiences that come into play in the Goffmanian approach. The examples of demonstrations proposed by Goffman in fact feature audiences made up of isolated, passive, interchangeable individuals. It will be useful to try to determine to what cases this model of the public applies.

THE AUDIENCES FOR DEMONSTRATIONS

If we want to pin down the extent to which public demonstrations can be apprehended as theatrical performances, we can look at the ways in which demonstrators and their intended audiences behave as "actors" and "spectators." In order to have a generic term that does not prejudge the precise nature of the intended recipients of the demonstrators' messages, I shall refer to these recipients as "demonstratees."

Various studies show that in many cases the roles of demonstrator and demonstratee are permeable. As we shall see, the activity of demonstrating sometimes corresponds to a full-time job. More often, though, it is an occasional occupation, in which an actor is led to switch between the roles of demonstrator and demonstratee in the course of his or her activities. This is the case in particular for scientists who produce public demonstrations and also attend those of their peers. Similarly, business people in various sectors of activity may give PowerPoint demonstrations for their colleagues and attend those given by others. In such cases, demonstrating is a professional skill more than anything else; it is not a full-time job.

In this context, individuals who adopt the role of demonstrator on a provisional basis may in effect become actors for the occasion. In my own work, I have seen researchers in artificial intelligence set aside their doubts about certain aspects of their projects for the purpose of making a public demonstration; they had to show confidence in order to try to get the financing they needed. There was therefore a discontinuity between the research centers, marked by the researcher's uncertainties, and the spaces where demonstrations were conducted. The former were partly constituted as backstage spaces and the latter as the main stages (Rosental 2007).

Another issue that needs to be addressed relates to the nature of the demonstratees. Are they always passive spectators, as they seem to be in Goffman's examples? And can we identify various types of demonstratees?

Certain audiences do indeed appear passive. We can observe this most notably in the context of scientific experiments demonstrated by way of mass media, which generally allow no direct interaction with the viewers (Collins 1988). Thus, the public for the experimental vaccination of livestock against anthrax, organized by Louis Pasteur on a farm in a French village in 1881, consisted primarily in readers of the available print media. The public's role was essentially to discover that one of Pasteur's predictions would prove accurate: after a certain date, all the vaccinated animals would survive, whereas the unvaccinated ones would die (Latour 1983, 150–152; Geison 1995).

But we find passive spectators in many other circumstances as well. For example, during religious processions in Cordova (Spain) in 2014, I was able to observe how public demonstrations of faith unfolded (see figure 1.2).[9] A clear separation was instituted between the actors participating in the processions and the witnesses. The demonstrators wore costumes and, in some cases, had undertaken lengthy preparations, while the spectators simply waited along the route. During certain processions, the actors adopted attitudes implying religious inspiration. They did not meet the eyes of the spectators, tourists, and members of the local population who were generally examining them in silence, confined to the role of inert observer. No participation, not even in the form of acclamation, could be seen. There were virtually no interactions between the actors plunged into apparent fervor, self-contained amid elaborate decorations, and the spectators, often casually dressed, looking for entertainment, emotion, and folklore.

Nevertheless, not all demonstratees can be characterized as essentially passive spectators or as observers limited to offering acclamation. They can be led to adopt varying attitudes during a single demonstration, and they may engage in interactions with the demonstrators. They may even be active participants from start to finish.

In the context of the high-tech demos that I have been able to study in the United States, for example, the demonstrators tended to start off with a monologue but then began to engage in dialogue with audience

1.2 Demonstration of faith during the Good Friday procession at Córdoba, Spain, in 2014. © 2014 Claude Rosental.

members, sometimes urging them to manipulate the tools of the demonstration themselves. In this configuration, the postures of demonstrator and demonstratee evolved in the course of the encounter (Rosental 2007). Here is the way a researcher participating in the development of software called Delta[10] assessed a demo of that software, in an email addressed to colleagues: "The demo with [these] folks went extremely well. They were able to use the system after a half hour demonstration, and threatened to commandeer my computer so they could play with it some more."

Similarly, during the demonstration of a software prototype in a laboratory at MIT, I observed the way the demonstratees were invited to express their reactions and their ideas, and to become advisors and potential partners of the project during the course of the interaction (Rosental 2019). This participatory status is clearly distinct both from that of silent observer and from that of simple consumer of entertainment.

In another instance of audience participation, during Microsoft's Tech-Days at the Palais des Congrès in Paris (February 10–12, 2015), I attended a demo of an "intelligent" whiteboard called Sensorit. This device made it possible to take data written by hand on the whiteboard and record it in electronic form. The demo had two parts. After spending a few minutes on an initial presentation of the device's capabilities, the demonstrator answered questions from the audience by supplying complementary demonstrations, structuring this part of his presentation around the questions raised.

Later we shall look in more detail at demos of this type made during Microsoft's TechDays. But we can already distinguish between "closed" demonstrations that do not anticipate interactions with the public and permit none, or almost none (these often involve products that are not designed to change in the short run), and "open" demonstrations that allow exchanges with the demonstratees and potentially leave margins of maneuver to transform the objects on display.[11] The demonstrations made during launchings of highly publicized products, like those of Steve Jobs, fall into the first category. Those that put into play the conception and sales of more personalized products, carried out before limited audiences, or mobilizing prototypes that might undergo significant modifications, generally fall into the second.

This distribution of demonstration types also turns up in the production of demos in the contemporary art world. Some of these put the public in the position of passive spectator, while others invite members of the public to interact with the demonstrator, or with the device that is programmed for that purpose. Juliette Sallin's interactive textile installation "Touch Me" offers a good illustration of the latter.[12] Designed to display the mediatory qualities of textiles, the demo was based on a carpet with small textile elements containing shape memory wires sewn onto its surface. When audience members touched the textile, electronic impulses were transmitted to the wires, raising the textile elements and making the carpet "move."

Clearly, then, demonstratees can be active participants on occasion. But we need to interrogate another of the characteristics Goffman attributes to them via his examples. To what extent must we view demonstratees as isolated, interchangeable individuals?

Various authors inspired by Goffmanian analyses have resorted to the notion of "awareness" to describe how individuals act in relation to what they observe in their immediate environment, even when they are concentrating on a specific task (Cahour and Pentimalli 2005). This notion is useful, for example, for describing how visitors to a museum explore what is on display in relation to the behavior of the people accompanying them and to that of visitors whom they do not know (Vom Lehn, Heath, and Hindmarsh 2001). This first step in nuancing Goffman's picture of demonstratees makes it possible to account for many instances of coordinated behaviors on the part of demonstratees.

For example, for a study I undertook during the Paris Fair in 2014, I was able to observe that the public consisted partly of groups of friends, couples, and families reacting collectively to the demonstrations, and interacting as these were taking place. Even as they followed the demonstrations, the demonstratees were observing the reactions of their companions and those of the unknown persons gathered around the stands. The demonstrators were generally quick to identify the group dynamics, especially those involving couples, and tried to take advantage of these dynamics during their demos.

One demonstrator explained to me how women, as he saw it, played a leading role in the evaluation and purchase of kitchen utensils, even though they generally sought the agreement of their spouses if the men were present. In addition to observations about gendered roles, the demonstrator also noted income-related and ethnic behaviors. For example, he considered middle-aged women from the French Caribbean to be ideal demonstratees. According to him, these women often started cascades of purchases by being the first to manifest a desire to acquire the products presented.

Testimony from a demonstrator at a fair in France, cited in a newspaper article, offers another illustration of recourse to this type of audience analysis and its use on the part of demonstrators (Anonymous 2013): "Gregory's method, used to convince the potential client: get him to participate. 'The skeptics roll up their sleeves and try out [the product] in front of everybody; it generally works out well. And the woman next to her husband encourages him in the task.'"

Although they are eminently visible, such phenomena have gotten very little attention from the sociologists who have studied the practices of demonstrators at fairs. These analysts have generally presented demonstratees as relatively homogeneous individuals who do not interact much with one another. The fact that audience members may have interpersonal connections even before the performance begins is rarely considered. This may be in part a result of the influence of the Goffmanian approach, which tends to stylize the actors by depriving them of numerous characteristics, most notably in terms of biography or social stratification.

We must not imagine that demonstrations at fairs are exceptional in this regard. According to one study, the members of the public present during presidential travels adopt behaviors that take the attitudes of others into account, and in particular the attitudes of the people accompanying them (Mariot 2001). In the context of street protests, I have observed, as have others, that media accounts were correlated with the political position of the reporters and their media outlets, and could be influenced by intermedia rivalries (Champagne 1984). Similarly, I observed in 2015 that the public for Microsoft's TechDays was not made up essentially of isolated, homogeneous individuals. It included, most notably, groups of students who had come together to ask company representatives about internships and to discover various innovations. There were also businessmen and engineers moving about in small groups with colleagues. Some women attendees brought gender identity to the fore, for example by reaching out via Twitter or responding to attempts to meet other female coders present at the conference.[13]

Studies of demos presented in business settings have produced similar results. Among other things, they bring to light audiences shaped by struggles taking place within organizations. One study shows how the audience for a demo presented by IT service providers in a company was structured from the outset around the divergent interests of its members (Smith 2009). The in-house IT engineers were opposed to a project led by an outside group. So were certain employees and managers, who favored the short-term profit of their unit over a new investment. The accountants and sales personnel had their own divergent views on the project being presented. The interests of the various subgroups played a decisive role in the way the demonstratees received the demo.

Such collective dynamics and the possibility that the audience for a demonstration will have been structured in advance do not necessarily exclude "audience construction" during the demonstration itself. This construction can take place on various levels. It can stem from the fact that individuals are brought together to form a more or less unprecedented and ephemeral group around a performance. It may be tied to the way the status of individuals evolves during the performance or at its end. During the performance, individuals may become passive "spectators," potentially active "witnesses," or active "participants," for instance, with assigned roles, rights, and obligations (for example, attending the performance without speaking and/or attesting subsequently to what was observed). At the end, the identities of the demonstratees may be modified in certain respects: for example, they may have become supporters of the project presented. In such a case, we have a public that "emerges" at least in part in reaction to the performance (Ezrahi 1990; Hilgartner 2000; Jasanoff 2005).

The public may also be "constructed" when the individuals who attend the performance have been selected in a more or less significant way. Demonstrators may have a wide range of motives and means for controlling access to their performances. For example, certain journalists presumed not to be hostile to a given political party may be permitted to attend a meeting that resembles a demonstration of strength. Visitors to a trade show may be filtered by the way invitations are issued or the way tickets are priced. The witnesses to an experiment carried out in public may be selected to ensure the credibility of the reports they produce, and to limit the risks of disturbing the social order by debates over their reliability (Simakova 2010; Coopmans 2011; Shapin 1988).[14]

Finally, we should note that an audience may be "constructed" when its identity is defined by the demonstrator in the script of the performance or during the performance itself. Here what is at stake is not the actual audience but the intended or ideal audience that the performance is, a priori, supposed to be addressing and may actually help bring into existence.[15]

A video recording of a demo posted on the website of the Danish health care system offers a good illustration.[16] This site is designed to convey information to the general public and to facilitate coordination among the actors in the national health system. The demo showed how an "ordinary citizen" who had just moved to a new city and was experiencing chest pains could use the website in his various interactions with health care providers. The patient was a man around forty years old wearing a suit and tie. He was presented as an actor in the management of his own body, not simply as a victim of his illnesses; on the contrary, he was portrayed as responsible for his own health and for the choices made in this arena. He entered into partnership with his doctors by giving them all relevant information and indicating what he wanted them to do. The demo then showed how the patient's personal data could be shared efficiently by all the actors in the health care system thanks to the workings of the website.

This demo simultaneously defined a political and economic project and an ideal audience. The latter was made up of "good citizens," suppliers and consumers of health services, responsible, educated individuals endowed with the ability to make the best possible use of the collective resources of the health care system. To the extent that this video was also used by its creators to promote the applications of the software in other sectors of activity, it simultaneously defined a second ideal public in other potential contexts, a public constituted by possible users of software that could coordinate the action of multiple actors. In other words, the demo was conceived to highlight a particular technology and to attract potential users whose identities, in particular as economic and political subjects, were multiple, and apt to evolve in relation to the new practices implied.

This reflection on the audiences for demonstrations is essential if we are to understand how demonstrations are conceived and carried out in practice. It is also indispensable for determining the extent to which public demonstrations may be understood as theatrical performances. As we have seen, the Goffmanian approach to demonstrations is limited in a number of respects. Highlighting these limits helps raise important issues.

First of all, assimilating these practices to technical redoings appears quite reductive. Public demonstrations are not limited to simulations of actions for utilitarian purposes, and they are not always conceived as such by the people who witness them. They appear just as apt to take the form of make-believe, ceremonials, or regroundings as to take the form of fabrications in Goffman's sense. Moreover, the idea that public demonstrations are agreed-upon fictions has to be significantly nuanced.

Contrasting situations have to be taken into account, especially nonuniform or more or less ambiguous presentations. Certain public demonstrations even resemble magicians' performances at times, and it is not always easy for the public to distinguish between "reality" and "illusion."

Furthermore, we have seen that the nature and behavior of audiences for demonstrations cannot be reduced to the simple configurations described in Goffman's *Frame Analysis*. Spectators are by no means always passive, isolated, and interchangeable individuals. Depending on the context, demonstratees may be more or less active, more or less united by preexisting bonds, and more or less endowed with personal characteristics that influence the unfolding of interactions. From this standpoint, the audiences for demonstrations are constructed by these events only in part.

If we want to grasp the nature of these practices with greater precision, many other issues need to be subjected to close examination. To what extent do public demonstrations entail constructed scenarios, stages, and manipulation of emotions, as the examples I have mentioned seem to suggest? I propose to approach these issues now by analyzing several detailed examples of how public demonstrations unfold.

2
PUBLIC DEMONSTRATIONS IN ACTION

To study how public demonstrations are conceived and carried out, I shall begin by describing a computer demo and a street demonstration in detail. Broadening the field of investigation, I shall go on to analyze the nature of the scenarios structuring these performances by comparing them to other performances (for example, demonstrations carried out in the banking sector, on television, or in the context of a fair). After considering the prescriptions of specialists in technology marketing who sketch out norms in this domain, I shall examine the role of the material devices brought to bear in various cases (scientific debates, street protests, architectural presentations, and PowerPoint sessions) in order to determine the extent to which demonstrations differ from simple oral presentations. I shall then look closely at the role played by the appeal to emotions in their use.

My analyses will thus bear in part on a type of demonstration that is characteristic of these practices according to Goffman, namely, demos. It is important to note that the ways in which demos can be carried out and the uses to which they can be put can be just as diverse and creative as those that characterize other forms, such as street protests.

As I indicated in the introduction, recourse to demos comes about in a multitude of situations that cannot be encapsulated by the highly publicized demos of Steve Jobs. It is possible to specify certain of their characteristics at the outset, in the light of my own investigations and those of others. For one thing, the size of the public may vary, from a single individual to a huge crowd. In addition, the forms of engagement with the public may be very diverse. Demos are used most notably by voluble vendors in marketplaces, by pugnacious sales representatives in the homes of their prospects, or by skilled presenters on television. They can be carried out by human beings or by machines programmed for the purpose. Certain devices—music synthesizers, for instance—can show off their capabilities in the store without human intervention, once the "demo" button is pressed.

A number of demos are carried out in familial or friendly contexts, whether to explain how a device works from a utilitarian perspective, to share enthusiasm for a newly acquired product, or to bring about sales, as in the context of a Tupperware party. They can also be presented in commercial contexts during events organized to launch new products, in high-tech trade shows,[1] in laboratories, in research seminars, and in many types of educational institutions. They come into play at different stages of technological projects, from initial sketches through exploration of various prototypes to presentation of the finished product.

In the context of research and development activities, demos may be carried out, for example, in the offices of venture capitalists, in lecture halls, or in research centers during guided visits designed for sponsors. Unlike demos for mass market products, these demos are often carried out by the actors who conceived of the products rather than by third parties (marketing specialists or sales personnel, for example). The demonstrators thus often link the features of the devices they are representing with the value of a theory or a method, or with the feasibility of a project. These demos are often recorded in video format and later disseminated more or less broadly, especially on the internet.

I was able to observe how a professor of control theory at MIT displayed the achievements of his team to frequent visitors during laboratory tours that consisted in a series of demos presenting a number of devices in motion (Rosental 2004). These tours were organized in stabilized forms down to the smallest detail. They were conceived especially for the industrial investors who were supporting the laboratory's research projects, or with whom the professor sought to sign new contracts. The

MIT researcher thus often played the role of business representative. He also presented demos of his projects in video form to investors when he traveled; thanks to this strategy, he was not obliged to take cumbersome or fragile equipment along. By concentrating on the production of a single bug-free performance, he achieved economies of scale in terms of the time devoted to making demos; in this way, he also escaped the possible mishaps of real-time replication.

Given the great diversity in the way demos, and public demonstrations more generally, are deployed, I need to reiterate that my analysis will by no means be exhaustive; I have simply selected cases for detailed study that I hope will constitute a useful point of departure for more extensive investigations.

FROM DEMOS TO STREET PROTESTS

It will be especially valuable to examine the unfolding of a demo produced by the MIT Media Lab, given that this laboratory is generally considered one of the leading sites for technology research in the United States, known for its large-scale use of demos, in keeping with its founder's slogan "demo or die."

When I undertook to carry out an ethnographic study in this facility, I noted that tours were frequently organized for industrialists. To this end, a number of glass partitions had been installed. This arrangement made it possible to show researchers at work with their equipment while minimizing disruptions. Supplementary displays of research projects were provided in demos prepared for the visitors (Rosental 2007). I have selected one of these to analyze in detail.

This demo, named Mediannotation,[2] was created in the context of a project whose development I followed over several months. The project was designed to create an iconic language and accompanying software that could be used to annotate multimedia documents; one of its goals was to enable journalists, media professionals, and others to find texts, images, and videos by formulating their searches with the help of concepts represented by specially designed icons. For example, one icon represented the presence of a video camera in a filmed sequence, while another signaled the fact that a person was looking at the camera. In a multimedia library holding documents annotated in this fashion, all sequences and images that included a video camera could in principle be located by selecting the corresponding icon.

The project brought together five researchers who devoted a great deal of time and energy to preparing and carrying out demos of their software before a variety of audiences (including colleagues, partners, and potential or actual sponsors). These demos were precious tools for the team members: they were used not only to persuade their interlocutors of the interest of the project, but also to gather critiques and useful advice that could influence the development of the project's content over time.

One demo in the form of a video clip presented some of the principles underlying the project. It also showed how to go about annotating a video with the help of one of the icons created. The transcription that follows is an excerpt:

[A man is talking in front of the Media Lab at MIT.] The challenge of representing video and other media is that we have to describe it at many different levels: Who's there? What's there? What's happening? . . . And the background that makes the event significant to us. For example, how do you describe the bit of video you are looking at right now? Well, it's winter, it's overcast, there is a bearded man, me, looking at the camera and talking. Behind me are some trees and a building. . . . The point of content description is retrieval, selecting the bit of video that you want to see. The point of retrieval is repurposing, using that segment in a new way. For example, this shot of the Media Lab will make a nice establishing shot for a video. OK, we wrecked that one! [The demonstrator is smiling.] But it shows something that happens on video and image all the time, the reuse of something that a camera team or photographer has already gone to the expense of making. The problem with that sort of reuse is that if you want to find the shot that we just shot, you need to be able to find it based on our description of it. One approach to this that we are exploring is a media annotation system . . . [Another demonstrator speaks in front of a computer, pointing to various parts of a screen and clicking on several icons.] We are trying to create a media representation of the content of video and audio. . . . The vertical axis is divided into tracks indicating what we can say about the video. Some of the information, we get automatically. . . . Other information needs to be annotated by the human user. Time: when something is occurring. Space: where. Important people, objects, actions, relations between people and objects . . . and any thoughts that the user wants to enter about the content. . . . Let us . . . annotate what [he] is doing. [The demonstrator is annotating the video that was just shot. The video appears in a window on the computer screen.] We go to the icons' pallet. It stores a set of iconic descriptors relevant to the scene that we are describing. Here again, we have time and space, characters, objects, camera moves, actions . . . and thoughts. Let us annotate what [he] is doing in the shot. So we take the icon for "looking" and annotate "[he] is looking at the camera." He is looking [The demonstrator is clicking on an icon and pasting it somewhere else], at the camera [The demonstrator is clicking on another icon and pasting it somewhere else]. He also happens to be talking in the shot. . . . We have the beginning of an iconic description of what [he] is doing in the shot. . . . A more difficult problem is finding the right icon among these thousands. Right now we have a browser which gives a sense of the space of icons and in which we can find the right icon fairly quickly.

Analyzing a demo on the basis of a transcription is an awkward exercise that confronts certain limits: some visual details and dynamics would require very lengthy descriptions to be made clear and analyzed fully. Still, certain structural characteristics can be highlighted. As in many other demos, this one takes as its starting point the presentation of a problem to which a technological apparatus is supposed to bring a solution. In the case in point, the software is presented as a simple answer to two general questions: how do you describe a video scene? And how do you make it locatable among others? Unable to allow exchanges with an actual audience, this demo in video form creates an ideal public. Two stages are set, moreover: one for the introductory monologue; the other for the central phase, when a video excerpt is annotated on a computer screen, and for the conclusion.

This demo unfolds in a structured way. It includes an introduction (presentation of the problem), a development section (presentation of the software's capabilities), and a conclusion (resolution of the problem with the help of an icon database). It brings together elements that are commonly found in this type of exercise: storytelling, recourse to humor (apt to incite a certain complicity with the viewers), and a manifest concern for efficiency and fluidity. In addition, it uses a concise exposition, tightly controlled timing designed to hold the viewers' attention from start to finish (the video lasts exactly 7 minutes 7 seconds), and a material apparatus that places the demonstration outside the realm of pure discourse. The ongoing interpretation that the demonstrator provides of his own actions in the second part of the presentation is another feature widely observable in demos. The speaker is producing a commentary designed to make sense of what is being presented: in particular, the objects visible on screen and the manipulations under way.

The demonstrators' visibility or invisibility is carefully orchestrated. The demonstrators appear at certain moments to present the actors behind the project, the problems raised, and some of the general concepts that are in play. They disappear during other sequences, so the viewers' attention can be focused on the workings of the apparatus being presented. In this way, the creators and their creation can be appreciated in turn.

To compare the structure of this demo with that of a quite different sort of demonstration, let us look at the Scientists on the March demonstration that I witnessed in Paris on October 17, 2014. Street protests unfold of course in a wide variety of ways; an analysis of this particular example will provide a first element for comparison.

This street demonstration brought together scientists, academics, and students from all over France to protest against the difficulties young researchers were having in finding stable employment, and also to denounce the inadequacy of the national government's investment in research. The demonstrators emphasized the danger of a brain drain and a decline in research in France, demanding that some of the funds allocated by the government to businesses for research and development (Crédit Impôt Recherche) be devoted to public institutions, state-funded research facilities, and institutions of higher education, in order to strengthen their capabilities and create stable jobs.

The demonstration had been organized by a researchers' organization called Scientists on the March, in collaboration with several faculty, student, and researchers' unions. It had been carefully prepared, with the participation of police officers in charge of security for the march. A number of demonstrators from the provinces had come by bicycle from all over the country to join the protest. These treks had also been well organized in advance, and well publicized. The production of the overall performance thus involved a large number of actors, including union members, representatives of civic associations, police officers, and journalists who would cover the event.

The demonstration had been carefully plotted out in spatial terms. A starting point and a terminus had been selected that would accommodate static crowds listening to introductory and concluding speeches. The itinerary had also been established in advance. The march progressed in a disciplined fashion, without overflowing the limits established by the police and the unions' security services. These latter ensured that the demonstrators would occupy only a single traffic lane and would enter intersections at precise moments, in order to limit the disruption to vehicular traffic and the risk of accidents.

The gathering was nonviolent in nature. The modes of protest consisted above all in speeches and slogans, whether these were expressed orally or in writing (chiefly on banners or placards). Nevertheless, a few police officers were placed at the entrances to streets leading to government ministries, in order to dissuade demonstrators from leaving the march to express their demands in front of government buildings. Overall, the police presence was relatively modest, and the officers were not highly equipped for repressive action. Moreover, the itinerary selected did not cross sensitive areas that might have led demonstrators to express their demands more vehemently or even violently.

The demonstration had also been planned with careful attention to timing. First of all, speeches were made before demonstrators and journalists gathered at the starting point. In this context, union and association leaders and researchers of all ages granted interviews to journalists from the various media, using the occasion to describe their working conditions and the purpose of their mobilization.

In a second phase, a cortege was formed and began to move. The participants' demands were expressed in various ways. Certain demonstrators, especially the younger ones, shouted their slogans out loud, sometimes accompanying them with songs and dances or strident whistles, and they sometimes staged sit-ins in the middle of the street. The demands were conveyed simultaneously by placards bearing sometimes dramatic or humorous formulations, such as "Sorry, Darwin, no evolution for you in research." The pace of the advancing cortege was regulated by the street crossings, which implied regular stops; the police controlled traffic at each intersection.

The demonstration ended with a final gathering marked by speeches, dancing, and ephemeral concerts. The moment of dispersal was itself timed, as illustrated by a brief altercation I witnessed. As the last demonstrators were leaving, the policeman reminded a lingering organization leader that it was time to go. The leader in turn expressed his dissatisfaction, retorting that it was, "after all," a demonstration. This remark conveyed a clear assertion that a street protest need not be so tightly orchestrated.

What audiences were actually targeted by this protest movement? The government was the primary target, since the goal of the march was to produce a change in budgetary policy. The public at large was invited to bear witness and encouraged to incite the government to act. The message addressed to these audiences was to be conveyed chiefly by the journalists spread throughout the crowd. As audience members, the journalists were relatively interactive, especially when it came to interviews with demonstrators. Media representatives welcomed their attention. In their introductory speeches, union and organization leaders had urged demonstrators to facilitate the journalists' work, especially that of the photojournalists; the speakers had emphasized that these journalists were contributing their support to the movement.

Pedestrians, tradespeople, and individuals passing by in their cars represented, in contrast, a more passive and less targeted public. In the Paris context, where demonstrations are frequent, many of these spectators witnessed the march with indifference. Others expressed their annoyance at the traffic jams it generated. Still others were amused by the spectacle, or showed curiosity about the demonstrators' goal. Astonished Asian visitors on tour buses captured the parade with their cameras, while a woman seated on a public bench expressed to another woman seated nearby her perplexity at the slogans on the banners: "Research? What's that?" Given that the demonstrators often moved along in small groups of colleagues, conversing among themselves and thus paying little attention to the passersby, interactions with these latter were fairly rare. They were essentially limited to brief exchanges with union representatives who were distributing flyers.

In comparing the unfolding of this demonstration with that of the demo analyzed earlier, a certain number of common points can be observed. In both cases, the public demonstration presents a problem and proposes a solution. Interpretations of the actions under way are supplied. The narrations that provide meaning appear quite elaborate. In the case of the street protest, a more detailed analysis would make it possible to bring to light the recourse to well-honed discourse, repertoires of highly stabilized narratives and incisive slogans. The same holds true for humor, which was used in both the demo and the street demonstration as a tool to create a certain complicity with the audience. Moreover, ideal publics can be clearly identified in each case.

In addition, the unfolding of each demonstration appears highly structured. The street protest can be seen as no more spontaneous and unorganized than the demo. Several stage settings were created in each case. In both cases we encounter extensive preparatory work, significant material and temporal calibration, and the mobilization of substantial equipment. These demonstrations seem to put to work scenarios as carefully wrought as those we find in most film and television productions (Chalvon-Demersay 1999; Grimaud 2003).

Before we go on to analyze additional dimensions of these two examples, it will be useful to ask, first, to what extent public demonstrations such as these are based on the use of scripts, and what sorts of scenarios can be brought into play.

SCRIPTS

To begin to answer these two questions, we shall focus first on how demos and demonstrations using software are structured. Later on, we shall look into other forms of public demonstrations, especially street protests. But in a first phase, I propose to analyze how famous demonstrators and communications experts in the realm of new technologies describe the construction of "good demos." The manner in which well-known demonstrators and communications experts in the realm of new technologies explain the practices and norms that go into the development of "good demos" will give us some initial clues for grasping their tenor, and for establishing some comparisons.[3]

While we know that technologies incorporate scripts for their own use,[4] is the same thing true of the demos used to present them? Let us start by considering the views of Guy Kawasaki, one of Apple's original marketing specialists and now an independent technology evangelist. As we have already seen, this expression typically characterizes an individual who is able to convert a large number of users to a technological product by adopting a stance comparable to that of certain Christian evangelists in the United States.[5] Kawasaki had the title of chief evangelist at Apple before he left to work on his own behalf and as a consultant for many other companies. Given his trajectory, his views and his skills in the area of high-tech demos correspond closely to those of the emblematic figure of technology evangelism at Apple, the late Steve Jobs.

For Kawasaki, a demo has to be simple, short, and well-paced. It must be based on a scenario that includes several elements: it must propose a comparison between the product presented and competing products; it must bring to light significant functional features of the product; and it must show clients what the technology in question can do for them (Gallo 2010).

In preparation for a major conference that he organized in 2016 under the title "Demo," during which start-up creators were to present six-minute demos of their products, Kawasaki offered eleven pieces of advice that he deemed indispensable to demonstrators applying to participate in the event and hoping to become "demo gods."[6] These demos had to be presented over a three-day period before venture capitalists, business analysts, marketing and communication specialists, and journalists covering the event. The declared objective was to put these different actors in contact with one another and to allow start-up creators to find funding, make sales, benefit from media coverage, and recruit staff.

Some of Kawasaki's advice dealt with the object of the demo (present only a "good product"), with the demo's material aspects (limit risk factors, for example by not using the internet, since network access might be unreliable), and with its preparation (improvise nothing; have duplicates for all the equipment in case of technical problems). Other recommendations concerned the conditions of presentation (carry out the demo single-handedly) and the explanations to present during its execution (avoid using technical jargon, so potential investors could reproduce the demonstrator's statements in speaking with their colleagues).

Still other advice bore more specifically on the way the performance should be organized. According to Kawasaki, demonstrators should plunge immediately into the demo without introducing themselves. They should show the most spectacular aspects of the technology right away rather than saving them for the end. Explanations about how the technology worked should be offered only in a second phase. Demonstrators should not answer questions from the audience until the demo was over. But the demo should end with the presentation of a striking feature of the technology, so the audience would be left with a powerful impression.

Somewhat comparable recommendations on how to develop a "good" scenario for a demo are offered by Carmine Gallo in *The Presentation Secrets of Steve Jobs* (Gallo 2010). This American communications consultant offers his advice to company executives and managers involved in launching technological products. In his book, Gallo analyzes the methods Steve Jobs used to prepare and carry out public performances (demos, PowerPoint presentations, speeches), and draws several lessons from them. His normative approach, much more developed than Kawasaki's, warrants close attention.

Gallo deems, first of all, that public performances should make liberal use of props and include demos. For him, several elements have to be included in demos' scripts. The scenario must be playful. It must engage the spectators' visual, kinesthetic, and auditory memory. It must also promote the creation of an emotional connection with the spectators, for, according to Gallo, emotions are useful for imprinting information on the viewer's or listener's memory. To this end, the scenario has to include surprises, has to create theatrical bombshells. These must be brought about with fluidity and at just the right moment, so that they do not "fall flat." Gallo recalls in this context the way Steve Jobs pulled a sample MacBook Air from a manila envelope during the launching of the new product, thus demonstrating in a memorable way how thin it was. The author also recalls how Steve Jobs had launched the Macintosh by letting the computer present itself.

According to Gallo, the scenario of a demo has to draw on examples and anecdotes. It should offer a better future in which the public can participate, as Steve Jobs did with zeal in his famous messianic approach. Moreover, the scenario needs to start with a question that everybody is wondering about, and it must answer a question that is essential for every user, namely, "What's in it for me?" On this point, Gallo makes it clear that Apple vendors are trained to explain to potential buyers in jargon-free terms how Apple products are going to simplify their lives and bring them functional features that matter to them.

Furthermore, Gallo contends that the scenario also has to include memorable "Twitter-style" slogans (which initially meant using no more than 140 characters). He saw the slogan used to launch the iPod in 2001, "1000 songs in your pocket," as an excellent example. Such slogans have to speak to the public, not to the demonstrator, and must be repeated. The language used in demos must be direct, simple, and telling. It has to be borrowed from everyday speech and include metaphors, comparisons, and analogies. It may appropriate familiar superlatives, as Steve Jobs did with terms such as "insanely great."[7] The point is to demonstrate a sense of humor and to communicate enthusiasm. Statistics, used in limited quantities, also have to be meaningful for the audience.

Any PowerPoint presentation included in a demonstration should consist essentially of photographs and other images, with few words. Bullet-point lists should be avoided, as should any statements that will be made in the oral presentation. Each slide should convey only a single message.

For Gallo, the scenario should designate an enemy (usually a commercial rival, as IBM was for Apple in 1984). This helps to mobilize an audience and to secure its fidelity to the brand as the "heroic figure" with which spectators can identify. The question of timing is also critical. Gallo explains how Steve Jobs succeeded in staging shows lasting ninety minutes without boring his audience. To achieve this, Jobs relied heavily on variety. He mixed in demos, videos, and success stories; he brought surprise guests on stage, produced testimonials from clients, experts, and partners (in person, in videos, or by quotation), and he did not fail to offer public thanks to his clients and his teams. According to Gallo, a video should last no more than three minutes, and a presentation fewer than ten. A demo should be as brief and as fluid as a movie trailer. On this point, Gallo mentions a demo by Steve Jobs that compared web browsers for mobile phones; it was less than three minutes long. Gallo also mentions a demo for GarageBand software that presented just one of its functions (Gallo 2010, 145).

Finally, according to Gallo, a demonstration should have no more than three parts and present no more than three characteristics of a product, owing to the limitations of the spectators' short-term memory. Moreover, the demonstration should tell a story that holds the audience's attention: it must be based on a well-constructed plot with interesting characters and visual effects like those of a Broadway play, and it must hold out some surprises for the end.

Before analyzing these diverse recommendations, it is useful to compare them to others formulated by professionals in the recording industry for the benefit of musicians who submit album proposals. These proposals typically include CDs, called demos, presenting musical excerpts. Here is sample advice found on a website intended for musical groups seeking to "make good demos."[8] In a context presented as highly competitive, and in which music producers have little time to examine proposals, the author recommends recording only three or four songs on the CD, putting the "best" one first (the one that has gotten the best reception from outsiders or in concerts), then the second-best, and so on. Moreover, musicians are advised to spend less than thirty seconds introducing the song's refrain. Finally, they must ensure that the recording is of good quality.

With this example, as with the previous ones, we can see how practitioners, consultants, and self-proclaimed experts are quick to produce normative judgments on the way demos should be structured. They go so far as to set quantitative limits for calibration (no more than x minutes, no more than y ideas, no more than z parts). Their advice converges on several points, such as keeping performances short and briskly paced, and beginning and ending with the most spectacular elements. Some recommendations (Gallo's in particular) are more specifically related to theater or film production in describing how to create scenarios for demos.

In my own studies of researchers working in artificial intelligence in the United States, the actors I met had no training at all in writing or staging scenarios. They had no communication coaches to advise them how to create good demos, and unlike Steve Jobs, they did not work in collaboration with professionals in the entertainment business. This helps explain why their demonstrations, as in the case of the Mediannotation project, were perceptibly out of sync with the norms I have just analyzed. However, the researchers had to deal with the need to produce demos of their achievements, either before an audience or in the form of video clips; thus they had to think about how to organize their presentations.

In this sense, they had to develop their own scenarios, although these were generally not written out, and even those that took written form lacked the precision that would usually have been required in the movie industry (Chalvon-Demersay 1999, Grimaud 2003). Thus they offered fewer handholds to a sociologist seeking to grasp them. To come up with a script for their demos, the researchers could use demos presented by their colleagues as models; they could rely on their own more or less extensive experience with this exercise; or they could follow the advice of more experienced researchers (the youngest among them could take this route). They also had models in movies and movie trailers that they had encountered in everyday life.

In addition, the researchers had specific skills related to the drafting of written demonstrations, and these skills generally influenced the structure of their demos. Indeed, many researchers complemented their scenarios with detailed reflections on the need for every sequence included, just as they reflected on the steps in demonstrations drafted for journal articles. In this sense, preparing demos usually was not an unwelcome exercise imposed essentially by funding constraints; it was more often an intellectually gratifying experience, a way of practicing science that was both enjoyable and stimulating. The "scripted" character of the demos was in keeping with the researchers' writing practices (Rosental 2007).

Under these conditions, the demos were often somewhat impervious to interruption, unlike written demonstrations. If the latter were designed to lead the reader step by step, from start to finish, the public demos guided the spectators' gaze even more strictly. Whereas readers of written texts can easily skim through one or more stages, can linger over a passage or go back to an earlier one, these demos required following a path marked by prepared exhibits and questions; they unfolded according to an implacable temporality that generally did not allow interruptions for questions.

As for their scenarios, one of their main objectives usually was to demonstrate the feasibility and the value of an approach or a technological project, as well as that of more general theses related to the theories at issue. The scenarios relied on the use of two or three senses on the part of the public: sight, hearing, and sometimes touch. As is the case with movie trailers, the designers of these demos typically sought to present the "best" of their achievements, conforming in this respect to the recommendations of communications consultants and experts that we have already observed. This convergence is also often manifested by a tight timetable. An excerpt from a message addressed by a member of the Delta project to colleagues illustrates this recurring preoccupation: "I'm going to see if I can speed up [this] problem a bit . . . 7 minutes is too long for a demo, and demos turn out to be important . . . 15 seconds is good. If you can do [that] as well in 15 seconds, then that would be 45 seconds total which would make a good demo."

The concern for not going on "too long" constrained demonstrators only in part in their development of the content of scenarios. These scenarios generally included an introduction and a conclusion, with a progression sometimes entailing a certain suspense. As in the case of Mediannotation, the scenarios could be organized to alternate between moments when the demonstrator was invisible, thus stressing the objective properties of what was on display, and moments when the demonstrator was in the limelight, in a position to create other effects (for example, humorous scenes intended to create a certain complicity with the viewers, or different framing angles to stimulate their attention). The demonstrators sometimes brought in a story, or used repetition to stress certain messages, and there were sometimes moments devoted to exchanges between the demonstrator and the public.

The progression of the demo could also be organized by a question-and-answer interplay modeled on Plato's dialogues. As in the case of the Mediannotation demo, the point was to use an organized series of questions to lead the public to adopt the conclusions sought by the demonstrator, typically by starting with current problems and showing how the technology made it possible to solve them. Constructing demos on this model was tantamount to building a utopia.

Among the elements generally considered for building scenarios, controlling attention was one. Like magicians preparing their feats of prestidigitation, demonstrators often sought, more or less deliberately, to direct the attention of the demonstratees toward certain aspects of their display—the strong points—and away from the weaker aspects (Rosental 2007; Jones 2011). They often tried to limit the risks of flaws—bugs—by choosing to present manipulations that had been thoroughly tested.

They sought to anticipate the expectations on the part of their audience, as well as possible objections, doubts, and potentially awkward questions. Thus the larger the public was expected to be, the more constraints there were on the scenarios.

Finally, the scenarios generally included examples of how to use the product on display, sometimes called "usage scenarios." They were intended for ideal or actual users, and according to the product they defined new tasks, new jobs, new behaviors, or new identities, as for example in the software developed for the Danish health care system we have already examined.

To complement this picture of the norms and practices encountered in the development of demonstration scenarios, let us look at another example of the use of such scenarios in sales practices. A study undertaken during the 2000s shows how the demonstration scenarios used by a Hungarian bank to advertise the advantageous terms of its lending practices evolved over the years (Vargha 2011).

The earliest demonstrations by bank counselors consisted in showing financing plans as a way for couples to realize their dreams. But with the liberalization of the banking market, the bank in question had to show above all that its offers were less costly than those of its competitors. The financial product proposed entailed a combination of savings plans that made clients eligible to take out loans at reduced rates, thanks to state subsidies, with loans proposed by partner banks (Vargha 2011, 224–225). For a while, the bank's counselors drew graphs by hand as they sat with their clients, before entering figures in a Microsoft Excel table to show the advantageous character of their product. Later, they acquired simulation software, which helped standardize their demonstrations. The script consisted in presenting graphs on screen and comparing the amounts of monthly payments according to the nature of the loans. It showed clients how their costs could be reduced if they opened several savings accounts. The demonstration culminated in a figure summarizing the gain that such a choice implied for the borrowers. The counselors then printed out three or four scenarios and gave the documents to potential clients with individualized annotations (230).

The available software tools could have allowed the counselors to skip most of the steps in the demonstration and go straight to the gains that would be accrued by opening several savings plans. But to get right to the point without going through the steps might have led the clients to question the accuracy of the calculations and arouse their suspicion. The bank's chosen scenario, based on a multistage demonstrative mechanism and on the display of intermediate results, was designed to make the final result look credible (Vargha 2011, 227). This element appeared all the more important in that the counselors involved did not always meet the description of a model ultra-competent demonstrator like Steve Jobs; they were not always capable of explaining why opening several savings accounts represented a source of economy for the clients (it actually had to do with an increase in state subsidies). Giving printouts to the clients was deemed a necessary step because the clients were often unable to remember the results and to reconstitute the elements in the demonstration once they were left on their own (227–228). The counselors tried to conclude the sale during their interaction (223), but when they did not succeed, they equipped their clients with further supports, thus prolonging the effects of face-to-face demonstrations.

Taking recourse to a variety of demonstrative tools is hardly an isolated practice, as we shall see. For now, it suffices to note that these sales demos put into play a carefully constructed script based on an elaborate progression and the use of comparisons. The quest for a certain simplicity, however, is in tension here with the risk that the demonstrator will not appear credible if he or she does not bring to bear a substantial demonstrative mechanism. Achieving a compromise between these two components, simplicity and substantiality, is thus a condition for the effectiveness of the script.

Moreover, we must recall that there is nothing exceptional about the use of a demonstrative scenario in the context of a commercial activity. To make this clear, let us examine the course of a demo carried out during a telemarketing show, a demo for "Basketonic" athletic shoes featured on a telemarketing program on the French channel M6 Boutique in 2012.[9] The script of this demo, which has several parts, includes elements that can be found in many other televised demos of the same type.

The script begins with the formulation of a problem: virtually everyone covers thousands of kilometers on foot, and this implies finding shoes that support good posture. Basketonic shoes are then presented as a solution to the problem, owing to the optimal height of their heels, 5.8 centimeters (about 2.3 inches). In contrast, according to the video, wearing shoes with higher or lower heels leads to the adoption of poor posture.

The demonstration of this assertion begins with the presentation of photos, comparative schemas, and computer graphics with commentary by the presenter. It continues with another "scientific" demonstration by a "physical therapist consultant." Wearing a white lab coat, this consultant supports his claims by displaying photos, charts, and a treatise on podology before the cameras. Testimonies from users filmed in homes and on the street extol the virtues of these sneakers. The presenter then adds that he has "repeated the experiment in the studio" with a model, showing the "results" obtained, which confirm his thesis. A further demonstrative element is added by the physical therapist, who points out details in the construction of the shoe's soles.

During the demonstration, other virtues of the shoes are mentioned, such as their elegance, their comfort, and the choice of colors. The cushioning they offer, the increased height of the person who wears them, and their capacity to create shapely legs and a slender silhouette are emphasized. Their low price completes the list of their virtues.

Images of two female models walking back and forth in the studio wearing Basketonics are shown, to reinforce the arguments. The demonstration ends with the presenter restating the thesis that has just been demonstrated.

This scenario appears comparable in many respects to those we have already examined, and in particular to those prescribed by Kawasaki and Gallo. Like Steve Jobs's demonstrations, this one relies on a series of varied interventions in a television studio, complemented by sequences filmed elsewhere. The demonstration is strategically paced, with no dead time, and it is short (three minutes forty-six seconds in all). It is based on the repetition of brief slogans such as "you become 5.8 centimeters taller." The scenario also brings comparisons into play, in this case with other types of shoes, stressing the unique features of the product on display. Moreover, it brings to light some of the functional qualities of the shoe, selected to appeal to the targeted users—especially women who would like to appear taller and thinner while wearing comfortable shoes.

Like the other demonstrations we have seen, it begins with a problem to which technology seems to offer a solution. Here is the transcription of the demo in full.

BASKETONIC DEMO

DEMONSTRATOR: [Close-up on the demonstrator holding a sneaker in his hand in a television studio.] In your life, you cover thousands and thousands of kilometers on foot, so you may as well do it with a good shoe, so you'll have good posture. I'm here to propose that you walk in our Basketonics. We're the only ones that make them. See this Basketonic: it's really elegant. You're going to enjoy walking. [Female models wearing Basketonics stroll across the studio.] It's available in two colors, beige or black. And it has a secret. Here, look closely and you'll see the secret of our Basketonic. [Animation showing a Basketonic.] This means that as soon as you put it on, you'll be automatically taller by 5.8 centimeters. Not only that: it will stretch out your silhouette, make it more shapely. And it will also improve your posture. [A triptych consisting of three photos showing a model barefoot or wearing shoes with different heel heights is presented on screen. Only wearing Basketonics is shown to support good posture.] You're going to gain, look, the equivalent of this woman's shoe with heels, with your Basketonic, but it won't show. [The demonstrator presents a shoe with high heels above a Basketonic, to show that they're the same height.] Joining me in the studio, here's our famous Jean Michel, our physical therapist consultant.

PHYSICAL THERAPIST: [Close-up on the therapist, holding a book in his hand.] Hi, Pierre.

DEMONSTRATOR: Hi, Jean Michel. [Close-up on the demonstrator.] So, why 5.8 centimeters? It's obvious, it makes you taller right away. But why not 6 or 7?

PHYSICAL THERAPIST: Because 5.8 centimeters, Pierre [a chart is projected on screen illustration showing the balanced distribution of body weight thanks to a heel height of 5.8 centimeters], simply makes the distribution of body weight equally balanced between the front and back of the foot. I didn't make this up, Pierre. [The therapist displays the book he has in his hands.] It's here in black and white in this little book I brought you [close-up image of the book]; it's a standard reference work for podiatrists. [Close-up of page 283, presenting a chart similar to the previous one.] And look, if I open it up to page 283, you have exactly the same argument, with this same angle. And see what happens. [Photo of a barefoot model adopting a crooked posture.] The distribution with bare feet is poor, and thanks to my Basketonics [photo of a model wearing Basketonics, adopting a straight posture], I'm going to have 50–50, perfect balance. And we can see it even better with the posture photos. [The camera is on the studio again, with the therapist showing the same triptych, pointing out the differences in posture.] Look at that, just what I told you, the body weight distribution is off. It's all on the back foot, and that person is going to be leaning completely backward and stressing her joints. On the other hand, if I have heels that are too high, what am I going to do? I'm going to move forward, but I'm going to have to bend to rebalance. Because people don't usually walk like that, fortunately! And thanks to my sneakers, look, perfect posture, the spine is exactly in line with the lower body.

DEMONSTRATOR: [The studio appears on screen again.] There, you're going to be able to walk thousands and thousands of kilometers [close-up on the demonstrator] with no problems with your Basketonics. [Close-up shot of Basketonics worn by a model.] You can pick them in beige or black. [Close-up of the demonstrator.] Let's listen to some folks who've already chosen them.

WITNESS 1: [A woman about sixty years old, seated in a living room. A subtitle indicates that she is "Alice Saiag, costume designer."] The advantage is that they're also super-attractive. And that works for me, because I can wear them with jeans, with leggings, to go shopping, walking, riding my bike.

WITNESS 2: [A man about thirty years old, seated in a living room; he is identified as "Olivier Ficco, a boilermaker in the plastics industry."] It's like they have little balls inside that massage your feet. And it's really something that makes you forget your shoes. You really feel better.

WITNESS 3: [A woman about thirty years old standing in a living room; she is "Anne-Frédérique Lamberdière, railway hostess."] Normally, when I go out, I have to wear heels to add a little height. [She is shown vacuuming her living room.] With these shoes, it doesn't show [close-up again of the woman standing in her living room], I gain a few centimeters, I feel thinner, more shapely.

WITNESS 2: [Close-up of the man seated in a living room.] I already work eight hours a day standing up [the man is shown walking in a street], all week long. And I always have a tendency to lean forward, to stay hunched over. [The man stops, waiting to cross the street. He looks somewhat hunched over.] And with this sole that's at an angle, I really have much straighter posture. [Close-up of the man seated in a living room.] And it's really very nice.

WITNESS 4: [A woman about 40 years old standing in a living room; she is "Basma Pierson, homemaker."]: What *I* love about these shoes is really that they're as high as a pair of heels, but as comfortable as a pair of sneakers.

DEMONSTRATOR: [The studio appears on screen again.] Right, our Basketonic [animation showing a Basketonic] makes you 5.8 centimeters taller right away, it's great, [photo of a barefoot model with crooked posture], it's really nice, with a shoe that's very elegant [comparison of two photos of the model, one with Basketonics, the other barefoot; charts and arrows stress the improved posture and increased height gained thanks to Basketonics], easy to wear. And we've tested them [close-up of the legs in the previous comparison, with a notation specifying that Basketonics "shape the legs"] again in the studio with our model [close-up of the demonstrator], Claire, who also gained 5.8 centimeters with these Basketonics. [Comparison of two photos showing different heights, about 1 meter 72 centimeters versus 1 meter 78 centimeters.] So to sum it up, Jean Michel [comparison of two photos, one showing the new model wearing Basketonics, the other barefoot], the advantages. Right, we've got it, you get taller, but also . . .

PHYSICAL THERAPIST: [The studio appears again on screen. While the therapist is delivering his explanations, two models stroll across the studio. In the foreground, three pairs of Basketonics are on display.] The advantages, thanks to that air-cushioned sole, on a cushion of air [close-up of the therapist manipulating a sole] you're going to absorb the shocks. And that's really important for your joints. [The therapist is shown explaining this to the demonstrator.] So you get into a good position with your back, and you shape your legs. [Close-up of Basketonics worn by a model.] What more could you ask, Pierre?

DEMONSTRATOR: Well, Basketonics, for walking thousands of kilometers [the studio appears on screen again: as the demonstrator speaks, the therapist is in the background and two models are strolling around the studio] with good posture. You can choose them [animation presenting a Basketonic] in beige or black. And one more thing [close-up of the demonstrator next to the therapist]: they're not expensive!

Several characteristics of the script of this demo can be found in other telemarketing scenarios that have been broadcast on French channels (TF1 and M6) and that I have analyzed in recent years: sequences filmed in studios and outside, series of explanations showing how a product works, testimonies from users and celebrities, and opinions from authoritative experts or scientists. The friendly, honest demeanor of the presenter, the objectivity and seriousness of specialists, and the seeming disinterestedness and enthusiasm of the users also represent tools mobilized to create trust. The demonstration thus incorporates a series of interventions designed to be mutually reinforcing. It is also arranged around its material components, which bring in the resources and skills associated with televised productions. Quantified data, graphs, and technological displays typically come in to support the discourse.

Moreover, posing a problem at the outset and demonstrating in what respect the technology being promoted constitutes a solution is a very common approach. The same is true of the foregrounding, via discourse and comparisons, of the purportedly unique, incomparable, exceptional character of the product presented.

Although close in many respects to televised demonstrations, those carried out in fairs are distinguished most notably by the fact that they rarely have access to the abundant and diversified resources of audiovisual production. In particular, they usually do not entail a sequence of scenes produced on multiple sites with many actors. Their scenarios are nevertheless carefully prepared. A demo for a cleaning product that I observed at a stand at the 2014 Paris Fair will illustrate this point.

This performance consisted in demonstrating the qualities of a multiuse degreasing liquid. On the stand, the audience could see an oven and a cooktop, several brand-name detergents, and containers of the product being sold. Various slogans on a green background pointed out that this product cut grease "with no scrubbing," that it was eco-certified, and that it had many uses, inside the house (kitchen, countertops, floors) and outside; it could even be used to wash cars. Endowed with a sonorous voice and the demeanor of an itinerant peddler with thirty years' experience, the salesman began his demonstration by raising a difficult problem: how to get rid of the grease that builds up on stovetops and oven racks after cooking.

He first showed how hard it was for familiar brand-name detergents to clean the grease from a stovetop burner. For this purpose, he rubbed a burner with a sponge saturated with one of those products. He stressed the damage done to the burner by over-vigorous scrubbing, and he mentioned the harm done to the environment by the classic products. In comparison, he showed how his degreasing liquid made it possible to clean the burner perfectly with a single swipe of the sponge, without effort and without risk to the equipment. The product was again presented as ecologically advantageous, and also as beneficial for the national workforce, because it was made in France.

The demonstrator completed his presentation by filling two glasses with a brownish, viscous liquid. When he poured a familiar brand-name detergent into one of the glasses, an emulsion appeared on the surface, but the color did not change. By contrast, the content of the second glass lost its color when he poured in a little degreasing liquid. Following this spectacular demonstration, he began to try to get members of the audience he had attracted to purchase his product.

The scenario for this demo was thus fairly simple, and it involved relatively few resources compared to those used in the script of the Basketonic demo. However, we find a similar path marked by the formulation of a problem to which technology clearly offered a solution, along with a progression entailing autonomous demonstrations aimed at mutual reinforcement (disappearance of grease from a surface with a swipe from a sponge, dilution of grease in a glass). The elements mobilized in the demos were not unique, either. The recourse to comparisons underlined the exceptional character of the product, just as the use of arguments in the form of slogans supplied an illustration.

2.1 Demonstration of a small household appliance at the Paris Fair in 2014. © 2014 Claude Rosental.

More generally, the scripts of the various demos I observed at the Paris Fair brought into play recurrent demonstrative figures (see figure 2.1). The scenarios were often based on recourse to comparisons between products and on posting their origins in order to stress their qualities (e.g., Germany and Switzerland for sturdiness and reliability, France for the defense of local employment). Foregrounding of media reports on the products (e.g., "as seen on TV") and of various certifications was frequently used to induce trust. Moreover, to try to create a certain complicity with the public, the demonstrators often punctuated their demos with humor, offered tastings, or communicated practical advice (e.g., recipes that involved the use of mixers). Certain demonstrations incorporated moments dedicated to audience participation. Spectators were thus invited to test vegetable peelers, for example, or special sponges for cleaning glass.

The existence of well-tested scripts went beyond the strict demo framework in the sense that the scripts took into account what came before and after the demo itself. Prior to the performance, an initial audience might be assembled via a pep talk, and at the end the demonstrator might announce a price and then offer multiple discounts and bonus products, to encourage sales in the immediate aftermath. Taken together, these demonstration-sales were the object of tightly calibrated timing; they commonly lasted ten to twenty minutes.

Studies focused on demos produced in fairs in France, the United States, and the United Kingdom show recourse to comparable scenarios, indicating that what I observed at the Paris Fair was in no way exceptional (Sherry 1998; Clark and Pinch 1995; Le Velly 2007). The convergences include the substantial use of equipment, by which I mean the entire set of "things" or "material supports," in Émile Durkheim's sense of the term, that are mobilized for a given demonstration to supplement the use of oral or written discourse or both (Durkheim [1897] 1951, 313–315). As several major figures in sociology have shown, especially Durkheim and Marcel Mauss, "material supports," generally speaking, play an essential role in social dynamics (Durkheim [1895] 1982, 71–72, 135–136; Mauss [1925] 1990). Many works in the sociology of science and technology have also stressed their importance, exploring in particular the roles played by "intermediary objects" in actions.[10] The use of equipment in public demonstrations is no exception to this rule, as we shall see.

EQUIPMENT

The notion of "barker" (French *bonimenteur*), commonly used to characterize demonstrators at fairs, is not generally welcomed by these actors (Anonymous 2013). It suggests a lack of honesty on their part, presenting them as clever liars, delivering blarney to drum up interest in their merchandise and find buyers. This expression introduces another hypothesis on the nature of their work, namely, that it is centered on argumentation and speech. Yet, as the foregoing analyses have just suggested, public demonstrations very often rely on gestures and involve expert manipulation of objects of all sorts. When these manipulations are not designed to speak for themselves, substituting for speech acts, they are articulated with verbal arguments in such a way as to shore them up. While they are essential elements of many demos, in various studies they are often overlooked in favor of a focus on the discursive dimension (Sherry 1998).

Other forms of public demonstrations follow a similar pattern. This is the case in particular for scientific demonstrations published in electronic forums, often called discussion groups. The expression "discussion group" is misleading in itself, for it implies interactions comparable to exchanges of oral arguments, whereas the computer technology used is an essential characteristic of these interactions. This technology has major repercussions on the way demonstrations are apprehended, and indeed on their very nature.

A study I undertook on a controversy over the demonstration of a theorem produced by a researcher in artificial intelligence named Charles Elkan made this gap very clear (Rosental 2008). The demonstration was initially presented during a major conference on artificial intelligence and published in the proceedings of the conference. A number of researchers challenged its validity shortly after its publication, for its conclusion called into question the foundations of fuzzy logic, one of the domains of artificial intelligence.

The earliest debates took place in an electronic forum that had been the theatre of numerous "summaries" and "presentations" of Elkan's theorem and its proof. It was not easy to get a copy of Elkan's article for some months after its publication, since the proceedings of the conference had not been widely disseminated right away. Although purporting to be faithful substitutes for the author's productions, the "summaries" and "presentations" posted on the forum were actually reformulations of the theorem and constructions of new demonstrative mechanisms. Often developed on the basis of secondary sources constituted by messages posted previously, these texts supported critiques or defenses of Elkan's demonstrations. To say what Elkan's theorem and its demonstration consisted in, or to play the role of judge by drafting a critique, were basically two forms of the same exercise; in both cases, it was a matter of producing new demonstrative texts.

Expressing viewpoints on the topic typically entailed taking up passages from earlier messages and inserting comments about a step in a demonstration or counterdemonstration that the writer deemed critical. This mode of writing ended up bringing out collective texts made up of several levels of citation and demonstrative linkages, but also prominent discordances. Some participants wrote that they did not see, at a given stage in the argument, what others showed or asserted that they had brought to light. Representations of Elkan's theorem and its demonstration were thus forged in part on the basis of these "substitutes," even if, according to the reactions expressed in the forum, the latter were read more or less attentively.

Moreover, to give more visibility to certain messages as compared with others, researchers used several different apparatuses (for example, some placed selected messages in a database accessible on the internet, or cited and reposted these messages frequently). Given the existence of these apparatuses, the texts posted on the forum did not all have the same impact on the formation of representations of the theorem and its demonstration. In the context of this material economy of access to the texts, the capacity of new demonstrations and counterdemonstrations to influence the representations of Elkan's original demonstration was actually quite variable.

The formation of these representations thus did not proceed from a summing-up of homogeneous individual examinations of an easily accessible text, obtained following attentive readings of that text. Given in particular the written dimension of the exchanges and the material economy involved in access to the texts, the process was not reducible to an exchange of ideas or to discussions comparable to confrontations of oral arguments. The representations of Elkan's theorem and his demonstration were primarily constituted in the course of interactions among readers through more or less visible texts. In other words, the viewpoints formed concerning the theorem were the results of readings of demonstrations that were public from the outset, and the material equipment of that publication played an essential role in the formation of representations of the actors and the nature of the demonstrations produced.

The importance of the material equipment can be observed in other forms of public demonstrations that are sometimes perceived as essentially discursive productions, comparable in that respect to academic exchanges. This is the case in particular for street protests, which are sometimes assimilated to the demands and the slogans they convey.

The case of the Scientists on the March event mentioned earlier offers a good illustration of the diversity of equipment necessary for this type of public demonstration. The equipment required included multiple

2.2 Scientists on the March street demonstration in Paris in 2014. © 2014 Claude Rosental.

supporting elements for displaying the demonstrators' slogans and their institutional affiliations (banners, placards, posters, balloons); costumes and accessories of all sorts (white lab coats for biologists, masks, academic gowns, or tools such as microscopes; see figure 2.2); sound systems (loudspeakers, microphones, amplifiers) to broadcast speeches and music; and musical instruments (such as whistles) used both to attract attention to the march and to give the event a festive character. Various vehicles were mobilized to transport the most cumbersome material. Finally, the police themselves had a good deal of equipment (e.g., vehicles, weapons, walkie-talkies).

These material elements played multiple roles. The diversity of institutional connections and academic dress displayed helped convey the sense that there was a massive mobilization of scientists, emphasizing the wide range of institutions, regions, and disciplines represented. Certain banners and the easily identifiable uniforms of police personnel were visual reference points for the demonstrators, serving to establish the physical limits of the march. The placards and loudspeakers allowed certain messages to emerge above the general commotion. The journalists present could thus summarize the demands of the demonstrators while showing animated scenes or colorful photos. Similarly, interviews broadcast on the radio could be punctuated with slogans from the demonstration.

To appreciate the importance of these apparently trivial material elements, it suffices to imagine a silent demonstration without visual supporting material or distinguishing clothing. The participants' demands and identities would be very hard to discern, and journalists would have far fewer resources for carrying out their task of producing an event intelligible to the public at large.

Moreover, if these elements are familiar to the point of appearing "natural" and thus easily overlooked by analysts, it is to the extent that the margins of maneuver for the production of a demonstration are limited. It is not easy for the organizers of a street demonstration to come up with audiovisual supports other than the ones already enumerated. It is understandable, therefore, that the same material elements will be used in one demonstration after another.

The importance of the equipment is easier to perceive in other forms of public demonstration. This is most notably the case for "technical presentations" in architecture. These exhibits take place in diverse contexts: for instance, architects may be competing in response to calls for proposals, they may be presenting more or less finished versions of projects to clients, or they may be submitting details to a local population for an urban planning project, seeking reactions or soliciting approval.

Many architects are fond of formulas that identify their style or their brand. They typically have no shortage of arguments and narratives to use in selling their projects or convincing their interlocutors to choose them. However, as several studies have shown, they also frequently use mock-ups, samples of building materials, plans and simulation software to carry out their demonstrations (Yaneva 2009; Stark and Paravel 2008; Houdart 2005).

In the creation of demonstrations, the development of sketches, scale models, computer-generated imagery, and spatial arrangements receives particular attention when it comes to preparing client visits to architectural firms, for example Rem Koolhaas's Office of Metropolitan Architecture in Rotterdam. It is not simply a matter of tools to facilitate visual perception by passive spectators. Purpose-built demonstrative elements can be manipulated by the public and give viewers a way to participate in the conception of a given project (Yaneva 2009).

Similarly, the analysis of a public demonstration carried out in Madrid by the architect Andres Ortega Perea suggests that this type of performance is not generally limited to speech animated by persuasive rhetoric. All sorts of devices were used in the Madrid presentation, some of them stemming from its typical spatial and temporal configuration. This configuration was very different from that of Koolhaas's clients' visits to his firm, during which the public could move about, manipulate the objects, and intervene at will. In Ortega Perea's demonstration, audience members were seated at a distance from the representations of the project and could speak only after the presentation was over (Yaneva 2009).

The competition among seven architects for the reconstruction at Ground Zero, the space left vacant in New York by the September 11 attacks, brought in a significant amount of demonstrative equipment. To show off their projects in a session open to the public, the candidates turned to PowerPoint presentations. The software allowed them to present images and animations using different types of materials. Their demonstrations were comparable, from this standpoint, to the telemarketing demo analyzed earlier, combining photos, diagrams, computer graphics, and other videos. Using the diverse capabilities of PowerPoint, the architects exploited effects of over-impression in order to make their building projects appear little by little on the empty space of Ground Zero. They sought to draw the viewers' attention to certain elements by using a blinking effect. They also used three-dimensional animations to conduct visits to their virtual building (Stark and Paravel 2008).

By these means, the architects sought to demonstrate the feasibility of their proposals, to make them tangible, and to win the public's favor. Their verbal arguments were thus combined with elements of visual rhetoric, based in part on the capabilities of the software used. In other words, the demonstrators drew on PowerPoint to carry out computer-assisted demonstrations, going beyond the order of discourse.

The case of PowerPoint software is interesting, because it illustrates the way in which a demonstration can be structured in part by the equipment it uses. PowerPoint presents a certain number of constraints and possibilities in terms of the path a demonstration can follow. In the

Ground Zero presentations, some of PowerPoint's particular capabilities were manifested in the way buildings were made to appear and their features highlighted. Similar effects can be observed in the production of various types of demonstrations within firms.

Studies have shown that public demonstrations using graphics and diagrams have multiplied within organizations since the 1970s (Gaglio 2009). While the earliest versions of PowerPoint made it possible to mix text with graphics, more advanced versions have allowed the addition of photos and videos. These diverse elements have become structuring components for the production of demonstrations within businesses, but also for running meetings and more generally in organizational life.

A study carried out in a French mobile phone company thus shows how PowerPoint presentations frame meetings within the company, through slide shows with commentary. These productions tend to elicit listening, note taking, reactions, and discussion (Gaglio 2009). The linked slides usually follow an orderly plan, with bullet points completed and adjusted orally, stories told about the pictures, usage scenarios displayed, and finally a recapitulation leading to discussion. This framing becomes a norm, a ritual, an outline for routine action, to such an extent that actors have difficulty doing things differently. PowerPoint files also represent supporting materials that can be transmitted to audience members at the end of meetings to compensate for deficits in attention; they can serve as memory aids and can be used to organize follow-up meetings.

PowerPoint uses are thus more diverse than those strictly related to the production of public demonstrations. But the extensive recourse to this software for demonstrative purposes, along with the multiplicity of other material elements used in the examples we have seen, raises an interesting question: have contemporary individuals come to view important limitations in (purely) oral demonstrations?

The potential of PowerPoint and its manifest utility as equipment for public demonstrations, along with the multiplicity of other material elements used in the examples we have seen, raises an interesting question: might demonstrators and demonstratees alike come to find that discourse alone no longer suffices for a successful demonstration? If we are to believe the ancient rhetorical treatises, the arguments of orators who master that art confer on them great power to persuade

(LaGrandeur 2003). My previous question might thus be reformulated this way: might the highly equipment-dependent character of contemporary public demonstrations result from the actors' recognition that the equipment-based approach is superior to an approach based strictly on verbal reasoning?

Some actors question that superiority: one study on the advertising campaign for a recent technological project, for example, showed that a marketing executive preferred well-constructed discourses to demos for the company's publicity campaign (Simakova 2010). It quickly became apparent, however, that this was a minority viewpoint; the decision to develop demos to launch the project was quickly imposed.

We also find examples of technologies for whose promotion oral presentations appear at first sight to be essential resources. But even in a simple case such as selling floor mops, it has become apparent that the use of demos can be preferred to a strictly oral argument, as I was able to observe in my studies at the Paris Fair. Similarly, when the technologies are of such imposing dimensions that their workings cannot be grasped on a single stage (for example, power plants), speeches evoking remote materials are often combined with demos focusing on specific parts of the devices (O'Neill and Nadaï 2012).

In short, demonstrations using material equipment seem to be preferred in many cases to purely oral performances. In these situations, the nature of the demonstrative medium is "impure": it combines gestures and discourse, showing and telling, objects and words. It is at the heart of what could be called a techno-rhetoric (Rosental 2018).

Another sort of "impurity" can be observed if we seek to contrast appeals to reason with appeals to emotion in the exercise of public demonstrations. Let us look into this question more closely.

APPEALS TO EMOTION

If we go back to the Aristotelian tradition, an effective argument must rely at once on *logos*, *pathos*, and *ethos*. A speech based on reason that plays on affects and is braced by the credibility of the orator can be expected, a priori, to win the adherence of its audience. By virtue of its emotional content, imagery, especially as conveyed by metaphors, must not be neglected in the mix of elements that constitute a persuasive argument (LaGrandeur 2003, 120).

The cases I have examined up to now appear to have incorporated such norms. The Mediannotation demo at MIT's Media Lab was based on an argument that could be characterized on various accounts as rational but that also relied on the seriousness of the demonstrators, the legitimacy of their institution, the humorous touches sprinkled throughout the narration, and the playful aspects of the images projected on a computer screen. The apparent robustness of the scientific and technological undertaking was not in contradiction with the relatively entertaining character of the performance, which sought to capture the attention of the spectators, to amuse them and arouse their enthusiasm.

Similarly, the demonstration of the merits of Basketonic shoes advanced multiple proofs and testimonies of their health benefits. The shoes were guaranteed by a medical professional citing his sources, by a presenter who looked honest, and by enthusiastic users. However, the demo also played on the desire that could be generated in some female television viewers by the prospect of discreetly growing a few centimeters taller while acquiring a slender figure, shapely legs, and elegant shoes: all this was stressed by the speaker's discourse and simultaneously illustrated by the models walking about in the studio.

The Scientists on the March street protest also featured solidly argued claims, expressed by researchers whose seriousness and sincerity were attested visually by their lab coats or academic garb. At the same time, the demonstration was designed to generate a certain empathy on the part of the general public, with young researchers attesting to their particularly precarious working conditions, and with individuals prepared to ride hundreds of miles on their bicycles to make themselves heard. Another goal of the event was to elicit fear of a brain drain, stressing the prospect that people who were contributing to the improvement of public health and living conditions in France might have to leave to work abroad. Furthermore, the demonstrators were trying to create a certain complicity on the part of the public at large by showing amusing slogans and a peaceful crowd moving along, singing and dancing, in a quasi-festive atmosphere.

These cases illustrate the way public demonstrations play on different types of affects, such as empathy, complicity, gaiety, enthusiasm, desire, fear, or hope. As I have been able to observe in several of my own studies, along with those of others, the nature of these affects can be quite diverse. The goal may be to move, to puzzle, or to surprise; to impress, to provoke, or to shock; to arouse indignation, to disgust, or to amuse; to charm, to frighten, or to invite dreaming. Let us recall, for example, how a demonstration of a financial product in Hungary consisted first of all in inviting clients to dream about owning their own homes (Vargha 2011). Another interesting case involves psychedelic demos produced by hackers that are designed to generate reactions of fear similar to the reactions triggered by horror films (Auray 1997). Countless examples could be invoked to illustrate the mobilization (or, conversely, the repression), in the course of public demonstrations, of a very large range of affects. But rather than drawing up a somewhat tedious list here, it will be more useful to inquire into the causes and effects of such mobilization. Let us start by looking at the views of communications experts and journalists on these questions.

A number of American technology evangelists and marketing consultants claim that "good" demos play on the emotions of the public (Gallo 2010). According to the communications consultant Carmine Gallo, Steve Jobs had the capacity to make his public dream during his demos. He was not simply selling computers; he was also selling the means for everyone to reveal his or her creative potential. Similarly, the CEO of Cisco, John Chambers, is not just selling routers during his demos, he is also promoting connections among people that could change their way of life.

Nevertheless, according to Gallo, a demo should create all sorts of emotional states beyond dreaming, for emotions leave a powerful mark on memory, a cardinal virtue for commercial communication. A public demonstration thus has to be surprising, has to include dramatic turns of events like those Steve Jobs was famous for producing.

For Gallo, the presenter must draw on acting abilities to generate all sorts of emotions. Demonstrators have to work on their diction, on the tone, volume, pacing, and silences that mark the various sequences of their oral presentation in order to generate enthusiasm and passion. To this end, they must add gestures to speech; they must use body language, establish continuous visual contact with the public, and adopt postures implying openness. They must avoid speaking in a monotone; they must show self-confidence and convey a sense of presence on stage in order to communicate their energy to their viewers, stimulate their imaginations, and make the demonstration enjoyable and engaging. According to Gallo, the demonstrator will arouse sympathetic feelings by smiling.

What is at stake in this process of stirring up emotions has often been noted by journalists. According to John Markoff (1996), "striking the spectators' imagination and their hearts and guts" is a critical factor for ensuring the success of a product launching. Steve Jobs's demonstration of the NeXT computer in San Francisco elicited similar comments in a newspaper article (Elmer-DeWitt 1988): "NeXT has captured the hearts and minds of some influential people. . . . In this business, making a computer with soul may be half the battle."

Certain authors in the area of innovation management have made similar analyses. For Joseph Lampel, for example, assessments of new technologies in the context of public demonstrations are not based on calculations alone. They are also based on emotional factors, the appeal to imagination in particular (Lampel 2001). There are usually many possible ways to appreciate the potential of a new technology. The evaluators retain only those elements that correspond to their expectations, their predictions, and the public's trust in the projects they observe around them. The affective elements influence this work of selection in a crucial way.

Demonstrations produced at fairs give rise to similar observations. One journalist who interviewed demonstrators at a fair in Metz, in France, and observed their practices summed up the results of his investigations as follows: "To sell a product in a demonstration is first of all to create a relaxed atmosphere, get the public to laugh. After that, the battle is won, most of the time. In any case, it's a success if the stand gets more and more crowded. The formulas, the phrasing, the gestures of the showman end up seducing the clients" (Anonymous 2013). Charming the public, generating sympathy, complicity, and a certain euphoria by showing humor and acting talents are presented here, in keeping with Gallo's prescriptions, as critical elements for ensuring the success of a demonstration. These elements echo the age-old reputation of itinerant peddlers as able to succeed in their demonstration-sales thanks to their didactic abilities, their power to seduce, and their aptitude to offer their audience the pleasure of a highly colorful spectacle (Braudel [1967–1979] 1992).[11] The ability to play on emotions is also often advanced as an essential dimension of the trade of demonstrator-vendor, whether it is practiced in stores or by traveling salesmen. Announcements of job openings for demonstrator-vendors published in France or in the United States show that this ability is a requirement, as the following examples make clear.

A recent listing for a job demonstrating food products in a supermarket in Illinois specified that candidates would have to show enthusiasm for their products, take pleasure in doing the work, and have a "sparkling" personality.[12] A similar call for personalities able to communicate enthusiasm was made in a recent announcement for a position as demonstrator-trainer in the dentistry sector in the Paris region.[13] A job offer published by a distributor of alcoholic beverages in France indicated the desire to recruit a demonstrator-vendor (male or female) who would be "dynamic, enthusiastic, with a gift for human relations."[14] In a slightly different register, an announcement listing a position as demonstrator for Ray-Ban sunglasses insisted that the candidates must have a "smiling" personality.[15]

These examples show the extent to which the ability to touch the public, to establish a certain complicity with an audience or to conquer its indifference are perceived as essential qualities for a demonstrator in the area of sales. Particularly important is the ability to transmit a certain enthusiasm to the demonstratees, to show them the possibility of escaping their ordinary routines for a moment, to arouse their curiosity and lead them to experience agreeable feelings, even pleasure.

Various sociological studies have shown in what ways emotions play an essential role in social life (Aranguren 2015; Déchaux 2015; Kalberg 2012, 291–300; Mauss 1921; Turner and Stets 2006), and in particular how widely they are used to strategic ends in everyday interactions (Clark 1997; Goffman 1961, 1967; Leroy 2008), or even in a logic of domination (Bourdieu [1980] 1990, [1998] 2002). More specifically, certain inquiries have brought to light the fact that a number of service professions expect a full commitment to emotional work on the part of their agents. The production and mastery of emotions thus appears to be an imperative in the exercise of many trades and for the establishment of a wide variety of commercial relations (Hochschild 1983). The examples I have just cited show that the practice of public demonstrations does not escape this "emotional imperative."

It is interesting to note that this imperative sometimes occupies the foreground, relegating so-called rational argument to second place. There is an echo of this phenomenon in Michel Audiard's 1974 film scenario *Comment réussir quand on est con et pleurnichard?*, released in English under the title *How to Do Well When You're a Jerk and a Crybaby*. Here we have a demonstrator-trade representative who does everything he can think of to sell alcoholic beverages to café owners. When his grandiloquent speeches, tastings, and offers of bonus products get him nowhere, he falls back on a quite different approach. He tries to inspire pity, especially from the café owners' wives, who turn out to be more sensitive than their husbands to this register. By lamenting his more or less made-up misfortunes in their presence, he manages to get them to put pressure on their spouses, and finally succeeds in selling his wares.

Here is a scenario in which we find a public that is not made up of isolated and interchangeable individuals, but one that is structured by preexisting relationships and whose members interact and adopt collective behaviors in relation to the demonstrator. However, to go back to the issue that is our main focus at this stage, the "emotional work" is disconnected, in the film's context—unlike the contexts we have examined previously—from the demonstration of the intrinsic virtues of the product in question. Moreover, this script features a limit case, the one in which the capacity to play on affects, and especially on pity rather than enthusiasm on the part of the audience, represents by far the most effective element for ensuring the success of the demonstration-sale.

If this configuration has a comic dimension, it is partly because it is somewhat out of phase with many real-life situations. Exhibiting misfortune and suffering to elicit compassion, whether or not it is accompanied by other sentiments such as indignation or anger, is a resource more often found in demonstrations by the homeless in a subway system, as happens in Paris, or in street protests for causes such as the struggle against torture by a given regime, or against world hunger. Anti-bullfighting demonstrations in France offer another example. Ethnographic observations have shown how images of the suffering and death of animals are chosen to arouse hostility toward this sort of practice without provoking disgust. Crossing that line would actually be counter-productive for the animal rights cause (Traïni 2010).

Street protests represent one type of public demonstration in the context of which play on affects is an easily perceptible and generally expected procedure, associated in many cases with the action of "leaders" able to galvanize crowds. A typical profile, to cite only this historical case, corresponds to Gustave Le Bon's descriptions of the apostle figure (Le Bon [1895] 2002). Apostles, in this sense, are individuals wholly imbued with the cause they are defending, able to fire up crowds by generating extreme emotions and getting them to carry out violent actions. To this end, apostles use statements that are at once simple and imagistic, formulated with a great deal of energy; these messages are then abundantly repeated and spread through a form of contagion. This profile of a man of action is close to that of the demagogue of classical Greece, according to Le Bon, but opposed to that of the "rhetor" who masters the art of argumentation but fails to be passionate about the cause being defended, which limits his power considerably.

Street protests are not the only type of public demonstration in which such apostle figures appear. Technology evangelists also have the reputation of "speaking from the gut" in order to set forth the prospect of a better future in which the public is invited to participate. According to Gallo (2010), Steve Jobs was animated by messianic zeal and sought to convert the public to the technologies he believed in, by carrying out demonstrations with a great deal of passion, enthusiasm, and energy.

Emotional work of this sort is not automatically crowned with success, however, and does not always act on the public in a homogenous way. Presupposing that the audience for a demo reacts in an undifferentiated fashion is, to be sure, a direct extension of the ideal and homogeneous representations of the potential users of a technology that often prevail in the process of innovation (Akrich, Callon, and Latour 2002a, 2002b). Similarly, it is all the easier to think that one is dealing with crowds unanimously inflamed by the words of a charismatic speaker if one imagines that these crowds are made of like-minded groups, belonging for example to the same social class.

In the context of the Scientists on the March demonstration, for example, the mobilizing speeches of the organizers certainly did not put the entire group of demonstrators into a trance. The demonstrators were by no means animated by extreme emotions; furthermore, they manifested a great diversity of attitudes. Similarly, my investigations at the Paris Fair brought to light a great diversity of reactions to the demonstrations of the most enthusiastic barkers. Several forms of distancing on the part of the public were observable, by way of ironic, prudent, or openly critical postures. Under these conditions, the spectators hardly appeared submerged in emotions to the point of losing their reasoning capacities.

Thus the response to emotional work on the part of demonstrators does not necessarily lie in the collective effervescence of the public. Foregrounding this type of phenomenon has long been a standard part of descriptive practices from Durkheim's work on (see Durkheim [1895] 1982), but without support from systematic studies. This type of foregrounding can be found in reports on various types of public meetings, over and beyond street protests. Accounts of presidential trips produced by journalists, for example, are apt to feature happy crowds, people abandoning their critical faculties collectively and uniformly in favor of strong emotions. But here too, careful investigations in fact reveal a great diversity of individual behaviors, ranging from detachment through conformity, timidity, self-mockery, mockery of others, or even exaggerated acclamation, to sincere engagement (Mariot 2001).

It is important to note, moreover, that the play on affects operated by demonstrators does not consist automatically in attempts to arouse extreme emotions on the basis of performances marked by a certain exuberance, outrage, or inauthenticity. In addition to some of the cases previously described, we can consider the unfolding of demos in the context of Microsoft's TechDays, which I observed in 2015. The tone of the demonstrators was often convivial, relatively sober but marked by subtle touches of humor. It did not veer toward exaggeration and extravagance but rather toward relaxation, soft persuasion, irony, and self-mockery. In many cases the demonstrators' attitudes were untheatrical, dispassionate, even marked with a certain detachment. It was often a question of sharing an interest in achievements by presenting curiosities rather than seeking to impress the public by spectacular displays.

This quest for a certain authenticity, or the absence of strategic simulation, can be compared to the attitudes that have been observed during various street protests (Traïni 2010). We can also compare it to the attitude of a doctoral candidate who participated in 2014 in a contest titled "My thesis in 180 seconds," organized in France by the National Center for Scientific Research and the Conference of University Presidents. This annual competition aims to select the doctoral candidate best able to demonstrate the interest of his or her thesis topic in the space of three minutes before a nonspecialist public.[16] The candidates who go through the multistep qualifying process benefit from the advice of actors and communication coaches in the preparation of their performances. The student in cellular biology who won a regional final in 2014 specified in an interview, however, that she had not been helped by such support. She said that she had stammered more than others and had made less use of "things like metaphors." She explained her success by the fact that she had been "spontaneous" and had made jokes appreciated by the public.[17]

Here again, we see that the prescriptions of the technology evangelists and communications consultants are by no means applied universally where demonstrations are concerned. The diversity of the modes in which public demonstrations are carried out can thus be observed just as much in the forms taken on by emotional work as in other aspects of demonstrations, such as the scenarios crafted or the equipment used. Still, we now have a better grasp of what structures demonstrative practices over and above their variations.

Like many other audiovisual productions, these practices often appear marked by a significant scripted component. The scenarios in play stem in many cases from a problem/solution schema and from a concern for the spectacular. The examples we have examined also bring to light a significant temporal and material calibration in demonstrations, a recourse to story-telling, comparisons, and slogans, as well as a desire to capture the attention of the public from start to finish, in order to lead the demonstratees in specific directions.

The equipment mobilized appears to be highly consequential, even determining. The public demonstrations observed differ as much from simple discourse as theatrical performances differ from the art of declamation. The cases we have seen bring to the fore dramatic scenes, a significant gestural repertoire, expert manipulations of diverse objects, and speech connected with these manipulations in a hermeneutic undertaking. Finally, emotional work appears as a basic, if not imperative, component of the demonstrative exercise.

If we are beginning to understand with more precision how these performances are put to work in concrete terms, questions remain about the preparations and the accompaniments they require. In what do these other tasks consist, and to what prescriptions do they give rise? These are the questions we shall take up next.

3

PREPARING DEMONSTRATIONS

Carmine Gallo, whose recommendations about the contents of successful demonstrations we have already seen, extended his advice to the preparations required for this type of performance. In his view, a demo must be prepared in great detail and rehearsed repeatedly. The first step consists in jotting down ideas for its content and then putting them in order. Next, content needs to be tried out on test audiences so as to improve it in response to their reaction. The smallest details—how the lighting is controlled, what to wear in a given situation, and so on—have to be thought through with care.

Once the script has been stabilized, the demonstrator must rehearse it for hours, so that the performance will be fluid and look "natural." For Gallo, these preparations must allow the demonstrator to do without notes and thus never lose eye contact with the public. At the very least, if notes are essential, they should be consulted only in quick glances, so the public will not notice, following the example of Steve Jobs, who sometimes glanced unobtrusively at the notes he had recorded with Keynote software. Demonstrators should make video recordings of some of their rehearsals and watch them with third parties, in order to note and purge any gestural or verbal "tics" and make sure the messages are always delivered energetically.

According to Gallo, the preparation must also entail anticipating possible questions and preparing answers. Similarly, demonstrators must anticipate potential bugs and come up with solutions in advance, so that such incidents will be as transparent as possible for the public. Demonstrators must train themselves to keep their cool, to avoid drawing attention to the incident: for example, by sharing an anecdote about bugs, explaining what the audience was supposed to have seen, then moving on.

If all these norms are inspired by the practices of Steve Jobs, we may well wonder to what extent they reflect the actual modes of preparation and organization utilized in the wide range of forms and contexts for public demonstrations that we are considering. To address that question, we shall look closely at some specific high-tech demos and a street protest, so that we can begin to see what "organizing" these performances means for the demonstrators, and what is at stake for them in their efforts. As we extend the field of inquiry, we shall inquire to what extent a dramaturgical approach to these activities is relevant: to what extent are we dealing with "stagings"? While we are asking when the preparations begin and what they may include, we can inquire more broadly into the nature of the various measures that may accompany a demonstration, before or after the actual performance, or both.

ORGANIZING AND REHEARSING

The history of science and technology is rife with instances of public experiments and demonstrations of machines resulting from lengthy preparation and abundant rehearsal on the part of demonstrators, with particularly good examples from seventeenth-century England (Schaffer 1994; Shapin 1984). The contemporary period has comparable examples in profusion.

For instance, in analyzing the development of the Delta software project in the United States, I was able to observe the considerable time and energy its creators expended on preparing "good" demos. They worked hard to find examples that would allow them to show the software in a favorable light, thus building up a stock of material—including repertoires of stabilized narratives—that they could use in order to avoid the pitfalls of improvising presentations. By working out an outline for their demos on the basis of a few well-mastered examples, they had the resources to foreground the strong points of their achievement; they could keep the audience from focusing on the unfinished or still unsatisfactory features of the software and had various means to control how viewers interpret its current capabilities. They could also anticipate questions and develop answers specifically applicable to the chosen examples.

Substantial organizational efforts may be observed in a large number of demos (Bloomfield and Vurdubakis 2002). In the days preceding the opening of Microsoft's TechDays conference in 2015, for example, a large number of tweets dealt with the considerable preparations being made by the demonstrators, whether in relation to their booths or in anticipation of the talks and thematic sessions to come. Illustrative photos were accompanied by messages such as "doing serious work on the last slides and demos in the speakers' room!"[1]

Such activity stems in part from demonstrators' frequent concern about avoiding bugs and crashes. In the Delta software development process, the creators thus preferred to use versions of their program that were not the most recent, but whose features and flaws they knew well because they had tested them at length. In this spirit, as one participant explained to a colleague in an email: "We are going to stop development of the current system this Friday, and 'freeze' it, we will only be using it for giving demos. . . . The new version will not be available for several months, and we are not planning on any intermediate versions that are robust enough to demo."

The process of stabilizing the Delta software in successive versions endowed with increasingly valuable features gave its creators a tool for limiting the risk that bugs would pop up during demos. But they also allowed demonstrators to put across the idea that their research could bring "concrete" results within a "reasonable" time period. Their promises appeared all the more solid in that a well-mastered material apparatus was available to shore them up. The prospect of conducting further research could be represented as an opportunity to add improvements to an already functional prototype with a precise set of specifications; this was a reassuring factor for actual or potential sponsors seeking certainty. Sponsors would hardly be likely to support a series of "experiments" that might lead to nothing at all; they needed to be persuaded to support the completion of a well-advanced or nearly finished project.

We have already seen this concern for avoiding technological mishaps in the advice on preparing demos given by technology evangelist Guy Kawasaki (Kawasaki and Moreno 2000). Other case studies illustrate the variety of measures taken by demonstrators to try to avoid technical problems: sticking to a stabilized script that has been intensively rehearsed; having more than one copy of the software on hand for a presentation; and when more than one creator might be able to explain the software's features, selecting the one who has best mastered the potential bugs to serve as demonstrator (Smith 2009).

To interpret this type of approach, we might be tempted to refer to work that posits the existence of forms of "techno-anxiety" in relation to the fear of demos crashing (Lunenfeld 2000). This kind of analysis has been produced most notably in the case of contemporary artists whose works rely on demonstrations of technologies. However, we must note that such crashes are sometimes actually sought and valorized. For example, Andrew Hertzfeld, a well-known demonstrator of Apple products, seemed to like having crashes happen during his demos, so he could show his audience that he was capable of overcoming them (Lunenfeld 2000). The idea was presumably to arouse admiration by displaying a certain form of virtuosity. Similarly, communities of hackers for whom demos constitute forms of entertainment and competition in the context of their meetings valorize crashes and the demonstrators' capacity to handle them. By contrast, demos without any slip-ups are deemed boring (Auray 1997).

Moreover, the emphasis on preparation for demos does not necessarily rule out improvisation. In many cases, demos are constructed at least in part on the fly, during the performance and during interactions with the public. Indeed, inopportune questions, unexpected responses on the part of the software, or moments devoted to audience interactions and technological manipulations represent opportunities for demonstrators to offer creative responses or adjustments to the scripts and explanations prepared in advance for the demos.

At the Microsoft TechDays in 2015, I witnessed some highly structured demos during major public lectures, but also others that unfolded with more flexibility and interaction at some of the booths, leaving a good

3.1 Demo at Microsoft's TechDays in Paris in 2015. © 2015 Claude Rosental.

deal of room for improvisation, even if they had been amply prepared and rehearsed at length. In some booths, exhibitors regularly proposed demonstrations of their products to individuals and to small groups of visitors that gathered for a few minutes before disbursing (see figure 3.1). The demonstrators ran through performances of variable lengths, given that the sessions were largely organized in response to the various questions formulated by members of the shifting audience.

The manipulations of technologies and the accompanying commentary thus represented sequences whose content appeared to be constituted in part in relation to interactions with the public, even if certain elements were repeated from one performance to the next. The reiterated explanations revealed the existence of repertoires of stabilized narratives on which the demonstrators could draw during interactions.

The excerpt that follows comes from a dialogue that took place during a demonstration of a 3D printer operating in a TechDays booth. It was launched by a visitor questioning the principal demonstrator; a second demonstrator intervened in the last few exchanges. During the dialogue, the printer was in the process of producing a small figurine.[2]

3D PRINTER DEMONSTRATION DURING TECHDAYS 2015

VISITOR: Hi. How does the printer work? How does it make things?

DEMONSTRATOR 1: It's plastic made from cornstarch. . . . This object was made all at once. It's printed as a flat shape, like this. That elephant, you might think, was printed in several parts, they've been stacked up. But no, it was made all at once. So, in the background, it's modelization that gives the printer orders . . . and then, these objects, it can take between two and three hours . . .

VISITOR: The printer adds material?

DEMONSTRATOR 1: Yes, it adds material and it builds up. So here, you have a little nozzle. You have the plastic coming in, it's heated at this stage. There's a little nozzle that makes plastic filaments, like this.

VISITOR: Do all 3D printers work like that?

DEMONSTRATOR 1: For this category, yes, that's the principle. It's adding material. . . . And in fact, it's super hot, and there's a little ventilation system . . . it gets up to 230 degrees [Celsius] . . .

VISITOR: And there are several price ranges?

DEMONSTRATOR 1: . . . This printer goes for €1,400.00. So they're starting to be completely affordable. . . . Lots of businesses today might find them interesting. And actually, it's super important.

VISITOR: What software do they use?

DEMONSTRATOR 1: There's just one software. . . . It's really the software that tells the nozzle what to do.

VISITOR: It's Microsoft software?

DEMONSTRATOR 1: No, no. . . . These printers are sold in Microsoft stores. . . . And we . . . have stories to tell. For example, I don't know if you know Kinect, the Kinect sensor that I have here. With Kinect, there's software called Kinect fusion. You scan, you have facial recognition. You get a digital file. And afterward, I print your head, like this.

VISITOR: Oh, really?

DEMONSTRATOR 1: So . . . we have lots of stories to tell. . . . There are lots of partners who have things they've developed with [Microsoft technologies], or that use [Microsoft technologies], or they don't. But here,

PREPARING DEMONSTRATIONS 85

it's to show a little bit of the innovative side.... Everything that's a connected object is very trendy, there you have it.

VISITOR: And how long does the printing take?

DEMONSTRATOR 1: Between two and three hours. But tomorrow morning, I'll try it. I'll start it off. Yep, that's what I said. I'm going to try to get there ...

VISITOR: Is the time it's supposed to take indicated?

DEMONSTRATOR 1: Here, it's at fifteen minutes. It's telling me that it'll take four minutes more. In four minutes, it'll be done.

VISITOR: So, you were saying that it's made of cornstarch?

DEMONSTRATOR 2: I think there's a bit of plastic mixed in, though. It's not 100 percent green, no!

DEMONSTRATOR 1: No, no, it's a mix.

DEMONSTRATOR 2: Right, it's recycled.

DEMONSTRATOR 1: It's ecological! It's organic! It's organic plastic! ...

VISITOR: You're surrounded by connected objects?

DEMONSTRATOR 1: Yes, completely. Over there, there are young people who've been working toward a competition in developing connected objects. They're showing their proto[type] to TechDay visitors, to test the idea a little. Over there [are] people from the Microsoft research department, the center of Microsoft research. They've come here, they've traveled a long way to show a little, uh, Microsoft researchers, what they're working on, and it's kind of the starting point for lots of things. So they're sort of Geo Find-It-All, over there. And over here [there are] lots of partners showing connected objects ...

VISITOR: Thanks very much.

DEMONSTRATOR 1: Sure, you're welcome.

These exchanges illustrate the way demonstrations could be carried out in the booths of TechDays around questions formulated by visitors. Somewhat like the old cabinets of curiosities, the salon brought together rare, unusual, and new objects; the regrouping helped emphasize the capacity for innovation of the company that organized the event. In the 3D printer booth, what was happening consisted as much in sharing enthusiasm or arousing surprise and interest as in highlighting the qualities of a product. The capacity to show and explain was coupled with the art of conversation. Nevertheless, the demonstration had several characteristics that we have already seen in less interactive demos. Intended for nonspecialists, it used very little technical vocabulary. Like Steve Jobs in his classic demos, the demonstrators relied on humor, a relaxed attitude, and a certain familiarity with their audience, along with a tendency to resort to storytelling.

The exchanges at the TechDays booths seemed to be structured, moreover, by routines of sociability that went beyond the specific context of the high-tech conference. Comparable phenomena have been observed and well analyzed in a study dealing with the presentation of demos featuring software for managing and correcting computer-assisted examinations (Capelle 2012). This program, developed by a small French company in Montpellier, was designed to scan examination copies—protecting their anonymity—and to make on-screen corrections with the help of special annotation tools. It also relieved human exam scorers of tedious tasks such as counting points and allowed them to communicate online with other scorers.

The development of this software was punctuated by frequent demos presenting different versions of the product prototype for clients or potential users. The analysis of various elements (actions taken, verbal exchanges, pace of delivery, facial expressions and body language, handling of physical space, production of written texts, manipulation of objects, presentation of visual elements) showed that the interactions that took place during demos were structured at elementary levels. In particular, the participants seem to be endowed with rights and duties that were manifested in the organization of turn-taking in speech. Their status evolved along with the roles adopted in the context of the exchanges. Thus the mode of regulating interactions in the demo framework was not fundamentally different from the one ethnomethodologists have observed in many other situations.[3]

In any case, it is worth noting that the organization of demos may have an iterative dimension. The development of a demo of SunSoft's NEO system offers a good illustration. Two engineers, Annette Wagner and Maria Capucciati, have attested to the way they helped develop a demo of this "application framework"—a set of software tools and components created to help developers design computer programs—even though these elements did not yet represent a functional apparatus (Wagner and Capucciati 1996). They were charged with showing programmers, during a major annual meeting of the SunSoft company, how the latter could use the architecture to build new applications. The demo was to be used as well by other actors within SunSoft, from participants in the NEO project to top managers, in a variety of arenas and for various audiences: clients, journalists, and assorted professionals in addition to computer technologists.

The development of this demonstration required significant preparation involving a large number of SunSoft actors and clients, and it entailed a large number of iterations. A team called DOVE, made up of programmers, interface creators, graphic designers, two "artistic directors" (Wagner and Capucciati) and a manager, was set up for the purpose. DOVE members interacted with the leaders of the group developing the NEO system and members of the marketing department as well as with clients of the company. In the course of their exchanges, Wagner and Capucciati tried to identify some "stories" that could serve as the basis for the demo, and they sought to settle on the visual details to be included, taking special care to prioritize images over texts. These "stories" consisted in narrations spelling out how NEO was going to solve concrete problems that clients were encountering, on the basis of clients' actual experiences. The artistic directors had the specific goal of highlighting the system's "credibility" and its quality, by presenting a scenario that could appear realistic to the target audiences and by ensuring the quality of the images and the appropriateness of the color choices made. They also had to ensure that the story line and the organization of the demo were satisfactory, offering their feedback to the other members of the DOVE team and collecting feedback from their test audiences.

Meetings with the assorted partners were organized for the purpose. Wagner and Capucciati recorded descriptions of the clients, including details on their companies and the business problems they encountered. Then they selected promising cases and helped develop possible scenarios for the demo in consultation with their various interlocutors. This work of construction and consultation was carried out with the help of hand-drawn sketches; computerized figures were used in a second phase, before the first programs for the demo were created. Various presentations and animations were tested, debates for and against took place, and the opinions of audiences, demonstrators, and programmers were analyzed systematically. The DOVE team had to serve as arbiter between opinions and demands deemed contradictory or inconsistent, especially those formulated by the NEO project leaders, who were consulted regularly. Since choices had to be made according to a precise timetable, certain paths that would have required a lot of programming time were abandoned.

The DOVE team finally settled on a case of software development for a large retail store chain to serve as the basis for the demo. This case featured a major retail chain that faced significant competition and needed to organize an effective sales promotion in response. The demo showed how NEO made it possible to solve practical problems that came up in situations such as calculating salaries of the sales force during promotional events.

The iterative approach continued through the demonstration that was presented during SunSoft's major conference designed for programmers. The evening before the performance, SunSoft leaders asked that softer colors be used, and an engineer proceeded to make the requested modifications without informing the graphic designers. According to Wagner and Capuciatti, the images projected during the demo turned out to lack sufficient contrast, and the colors were too pale. The lesson the artistic directors drew from this was that the team should avoid making last-minute changes without testing them first, and the problem was corrected for subsequent demos. Moreover, because Wagner and Capucciati were seated in the audience in order to analyze audience reactions to the demo, they were able to observe how certain spectators had followed the presentation closely and had discussed some concrete prospects that the system could offer them.

This case shows how the work of analyzing public reactions in the context of preparing and presenting demos can be carried out in a continuous and systematic way. However, it also shows how the preparation of a demo can involve a consistent set of operations and an iterative construction based on a large group working together with a clear distribution of roles, stemming from an overall project management approach.

A certain number of elements and prescriptions related to demo development that we have already examined reappear here, such as the construction of a scenario, the effort to tell a story, and the attention to the visual and material character of the demonstration. We also find here once again the recourse to test audiences, to "robust" versions of the devices used for demonstration purposes, and the formulation of problems encountered by audience members, followed by efforts to show how the new technology could offer them solutions. The specific case we have been examining provides, in addition, the illustration of a demo featuring a fictional device—a "vaporware product"—that cannot be manipulated by the public.

It is interesting to compare these preparations with those that can be put into play in the context of a street protest. The example of the Scientists on the March public demonstration is instructive in this regard. This gathering was the result of a long preparatory effort. It included, first of all, the orchestration of bicycle routes originating in various French cities so that the riders would arrive simultaneously on the day of the event in Paris; some riders would spend as many as three weeks on the road. Regional organizing committees had been set up for this purpose. Dedicated web pages provided the participants with a great deal of practical information (meeting places, sign-up forms, itineraries, and so on). The websites included appeals for funds to help support the cyclists and instructions for helping the latter find free housing along the way; they also reproduced regional newspaper articles about the adventure.

A major logistical effort was also required to make the march in the city of Paris possible. A very large number of actors would be involved (unions, scientific associations, journalists, committees focused on specific themes, local organizing committees from universities and research associations, and so on); information had to be shared and activities coordinated among all of these. Considerable efforts were also required for the mobilization of various material resources (trucks and sound systems, for example), arrangements for security services, and communications. These latter included the development of slogans, banners, posters, logos, and websites. They entailed collecting support from elected officials and various organizations, scheduling interviews with journalists, and drafting press releases that were then systematically put online.[4] They also presupposed the creation of entertaining material—chiefly in the form of cartoons and a support program called "adopt a researcher"—that would explain the motives behind the movement and the importance of research to the public at large.

In this very different type of public demonstration we find the same significantly collective dimension that we have seen in the preparation of certain demos, even if this street protest was organized on a much broader scale. The attention to the material details of the demonstration is equally noticeable. The work on slogans and logos for the march parallels the work put into graphics and narrations during the creation of many demos. The standard concern for avoiding bugs and crashes during public demonstrations of technology has its equivalent in the standard preoccupation among both police forces and event organizers to avoid spillovers and violence, the latter having often been identified as detrimental to the causes being defended (Champagne 1984; Fillieule 1999).

The definition of the path of a street protest is viewed in this regard as a preventive measure just as important as the definition of a well-mastered version of a prototype to be used in a demo. Several studies show that establishing the itinerary of a street protest can require significant preparatory efforts, including substantial negotiations between the organizers and the police, especially in France (Fillieule and Jobard 1998). Keeping the marchers at a considerable remove from government buildings is one of the tools regularly used to limit the risk of confrontation. The course authorized for Scientists on the March thus placed the demonstrators at a significant distance from the institutions of centralized power.

The organizational work behind many major street protests in France can actually be described as the coproduction of an event with the police. The latter's efforts appear to be oriented toward management of the various kinds of disturbances to "public order" that range from violence against persons and property to traffic problems. These efforts rely on a number of skills, techniques, and procedures that vary from country to country and from one legal framework to another (Fillieule and Jobard 1998). Depending on the circumstances, they may entail negotiations with the organizers, collaborations with internal security services, and the formulation of orders governing the conduct of the demonstration (itinerary, timing, modes of crowd control, and so on).[5] The role of the police forces in the organization of certain demonstrations is in fact as important as that of the marketing services of large companies in the preparation of certain demos.

Still, we need to note that this collective production of public demonstrations arises in many cases from a more specific work of staging. Let us look more closely at what this entails.

STAGING

As the foregoing discussion suggests, public demonstrations often entail elaborate stagings. The term "staging" refers here to the creation of a stage, that is, a separate, furnished space in which actors are to act in an expected, organized way and are to be observed by others. The work of staging public demonstrations is easy to grasp in the context of an event such as Scientists on the March. First of all, as we have seen, that street protest was preceded by highly publicized bicycle treks to Paris from many cities in France. These itineraries represented a first set of scenes bringing together cyclists being observed and interviewed by journalists from regional print media, and they contributed to the staging of the principal demonstration to come. They were more evocative of pilgrimages than of athletic exploits, although the inspiration was secular. These stages presented actors who were victims, or supporters of victims, of the precarious economic situation of young researchers; the actors spared no effort to offer a vigorous defense of what they saw as a just cause. The organizers sought to elicit admiration, to the extent that the cyclists were traversing long distances at the cost of some physical hardship, in order to express their indignation.

The Paris street protest, for its part, created a temporary and mobile stage in the urban space as it progressed, a stage on which journalists and passersby alike could observe and film actions that were just as premeditated and organized as those of the cyclists on their way to Paris. This stage was especially arranged to allow the unfolding of a demonstration of strength.[6] In fact, the march put on display the presence of numerous clusters gathered behind banners, flags, and placards. The mass of demonstrators could thus be observed in groupings according to their declared memberships in various academic organizations, unions, associations, and disciplines (biology, physics, and so on). These groupings were reinforced by diversity in dress (lab coats, academic regalia) that could further differentiate among research specializations, trades, and university affiliations. The march thus took on the air of a Noah's Ark, bringing together in a visible way all "species" of scientists. This multiplication of distinctive signs made it possible to highlight strong territorial and institutional solidarity and the broadly representative character of the movement. Combined with the spread of the march over a considerable distance, the visibly significant number of demonstrators, and the presence of well-known political figures and elected officials wearing their tricolor sashes, the staging helped demonstrate the strength of the mobilization.

While the shifting space of the gathering allowed interactions between demonstrators and journalists in the form of interviews, it also gave journalists a place to observe various visual elements and action sequences: décors, dress, animations, and so on. To the extent that it provided something to see, the entire event could be characterized as a space for a show or spectacle, allowing media representatives to carry out their work of making news.

The creation of stagings can be observed in many other types of demonstrations. As an example, we can look back at the street protest of the National Federation of Farmers' Unions that took place in Paris in 1982. The representative nature of the organization was on display in a march that brought together groups of farmers from various regions, identifiable by their banners and their traditional garb (Champagne 1984). Similarly, the organizers of a number of anti-bullfighting demonstrations that took place in France in the 2000s anticipated scrutiny from journalists by paying attention to crowd control and dynamics as well as to messaging and signage (Traïni 2010). In an analogous situation in the United Kingdom, organizers of protests against the construction of a new road in Newbury prepared stagings for journalists: in front of the cameras, demonstrators disguised as animals perched in trees taunted the police personnel who were trying to dislodge them (Hindle 2006; Barry 2001). Some humorous stagings were also developed starting in the 1980s in street protests organized in the context of queer movements (Shepard 2009).

The creation of stagings for street protests can be compared to the process used for demos. The case of the demo for the NEO system is enlightening in this respect. We have seen how the two "artistic directors" contributed to constructing a "realistic" scenario around the problems encountered by a large retail store chain; they offered advice about the production of the demo and paid special attention to the choice of images and colors (Wagner and Capucciati 1996). In their paper, the two engineers represented themselves as directors responsible for making sure that the plot of the scenario was effective and that the action progressed smoothly; they also depicted themselves as stage managers and lighting technicians in charge of setting up and adjusting the visual effects. The authors spelled out the respects in which they had sought to construct a stage on which "the invisible" could appear, referring explicitly to the conceptions of the English theater director Peter Brook (Brook 1968). In the NEO case, the goal was to bring to light "how the business process objects were working with each other and with the other layers of the system" (Wagner and Capucciati 1996, 5).

The notion of "staging" thus seemed well adapted to the NEO case. The process of constituting and organizing a space and specific actions and presenting them to diverse audiences resembles the preparatory process for the street protests described previously. However, it bears an even closer resemblance to the process involved in theatrical activities, both in its inspiration and in the specific modalities of its unfolding.

With the Delta project in the United States, we have already seen a case in which the development of software led to the creation of stage settings to be used in demos. Throughout the process, and especially in the context of their demos, the researchers and engineers involved sought to limit the doubts of potential users as to the future capabilities of the software. The places where Delta's demonstrations were presented thus constituted stage settings dedicated to the observation of technological achievements by various audiences, a process quite distinct from those involved in the development of such performances. Two types of spaces were thus constructed simultaneously: stage settings where the demonstrations were manufactured, marked by important but scarcely visible preparations, on the one hand, and settings for the performance of demonstrations, on the other.

This work of staging was manifested above all in the highlighting of the software's capabilities. The Delta software was intended to be used to determine the possible trajectories of spacecrafts. One of the creators' challenges was to be able to make clear the value of using this new software as opposed to the existing tools available. The demonstrators needed to be able to show that Delta allowed its users to solve the problems of identifying the possible trajectories more easily and more rapidly than the then-current procedures allowed.

To that end, a series of problems had been put to the test on the earliest prototypes of the software, and some fifteen cases had been selected to illustrate the program's effectiveness. The first demos consisted in submitting some of these problems to the Delta software sequentially, to show how easy it was to formulate the problems thanks to the program's interface, and to emphasize how quickly solutions were found.

The demo's creators had attended to numerous details: for example, they had decided at one point not to print the visible results on screen during the demos, so that the time required to solve the problems would appear as short as possible. Printing out the solutions took quite a few seconds, and would thus slow down the demo; concerned about efficiency, its authors made sure that it would not last more than a few minutes.

In this case, we have seen once again how specific actions intended to be spectacular had been carefully prepared and orchestrated to be exhibited in front of third parties. And we can find many other examples of demos where the sophistication of the staging is unmistakable. The Mediannotation demo we have already examined offers another good illustration. Let us recall that that demo included several scenes. In the first, filmed outdoors, one of the creators evoked in concrete terms the problem of how to describe a video sequence. The second showed how a media professional confronting that problem could resolve it by adopting a system of annotation by means of icons. A second participant in the project was then filmed in the process of showing, on a computer screen, how to describe a video sequence with the help of Mediannotation.

The demo for Basketonic shoes was staged even more elaborately. It brought into play a larger number of actors (presenters, models, a physical therapist, users offering testimonials), and material resources (photos, animations, charts, graphs, computer graphics, and various material objects). It strung together in a skillfully ordered and carefully paced progression multiple scenes filmed outdoors and indoors (principally in a television studio and in users' living rooms). These scenes were marked by various actions: testimonies, experiments, explanations, manipulations, and episodes of strolling about. The décor and the lighting as well as the acting were quite sophisticated. Moreover, the notion of staging is especially relevant in this case, because the demo drew on the skills and competencies characteristically involved in a televised production.

Yet another set of practices can be viewed in the same light, namely, the practices of the technology evangelists and consultants in technology marketing whose prescriptions we have already examined. References to the theater and to the work of staging were quite explicit in those prescriptions, and the competencies of the actors in these fields were notable. Carmine Gallo, for example, took his inspiration from Broadway plays (2010), and approached the task of the demonstrators as if they were playing theatrical roles. He insisted on the usefulness of having guests who could join the demonstrator on stage, of having a well-constructed plot and carefully planned visual effects. He also stressed the importance of rehearsals and of lighting design. Steve Jobs worked in collaboration with theater professionals, and Jim Grubb, Cisco's chief demonstration officer, had been trained in the dramatic arts.

A comparable effort in staging can be observed as well in the context of high-tech showcases. One study dealing with the modalities of participation of a company developing radio frequency identification products (RFID) at a high-tech conference offers a good illustration (Simakova 2010). Members of the marketing department of that company paid very close attention to the conditions in which they were going to set up their booth. They had studied in detail the elements of the décor, such as the color of the carpeting, the arrangement of the walls, and the content of the posters. They were concerned about whether the company's logo would be visible from the main entrance to the salon, as well as about their booth's position relative to those of partners or rival enterprises. A marketing specialist from the company, deeming certain booths "dull," wanted theirs to be more like a "theatre": as models, he took one inspired by Disney motifs, and another that allowed visitors to operate a toy train. The way the technology should be presented had also been worked out carefully, not only for the exhibits in the stand but also for the talks that would be given in the context of the event. Not all exhibitors were offering live demos, moreover. Some used video clips prepared in advance to explain how their technology worked.

The use of the term "theatrical staging" thus seems well suited to a wide range of cases. The term does not serve solely as a metaphor; it refers in some cases to the possession of skills and competencies manifested by theater and television professionals, the same skills and competencies advocated by certain specialists in marketing and communication. These dynamics also justify the notion of "theater of use"; this term is sometimes used more broadly to designate the way a technological system can be presented in a spectacular fashion through the actions of a demonstrator interacting with the technology (Smith 2009).

We must note that the modes of grasping many other types of public demonstrations also emphasize the relevance of the concept of theatrical staging. Itinerant peddlers used to characterize their bargain-price sales as "comedies" (Duval 1981). Many attorneys practicing criminal law in France perceive their courtroom pleas as a form of theatrical activity, marked especially by flamboyant gestures.[7] The adoption of a dramatic approach to analyze the course of the highly publicized Oliver North hearings before the United States Senate in 1987 can be justified on several levels (Lynch and Bogen 1996). Beyond the judicial universe, the public demonstrations that mark debates around the reports of experts within the American Academy of Sciences can also usefully be described in terms of theatrical performances (Hilgartner 2000).

Still, in many cases the notion of theatrical staging has its limitations and must be used with discretion.[8] In the context of Microsoft's TechDays, for example, we saw how some demonstrators primarily tried to share their interest in specific technologies without seeking to work out well-developed stagings. They favored a relaxed, straightforward approach that allowed for improvisation, in contrast to strategic simulations requiring lengthy preparation. Some demonstrators were employees of big companies, and their personal stakes in a successful demonstration were limited. They were not in charge of start-ups, did not have the pressure of ensuring the survival of small companies, and, generally speaking, did not have to worry about performance evaluations by their employers. For these exhibitors, the pleasure they might take in presenting and explaining how some astonishing product worked, and in interacting in a friendly way with visitors, could take precedence over business concerns.

Another example can illustrate why the notion of theatrical staging must be used with caution to account for the preparation and performance of public demonstrations in all their diversity. It has to do with the controversies that arose around the adaptation of Arabo-Andalusian music for programming in France (Jaffré 2005). For certain organizers of cultural events, this music was undeniably material fit for presentation in entertaining spectacles. In contrast, for certain ethnomusicologists, the works in question constituted above all proofs of the existence of a culture that could not be staged and grasped as objects for public entertainment. A case like this shows that the notion of theatrical staging does not represent merely an analytical resource. It may in some cases represent a problematic category for actors and may need to be treated as such.

However, the work of preparing and presenting public demonstrations raises additional questions; we may wonder, for example, just when this work begins and when it is finally complete. Let us now try to see to what extent this pre-demonstration effort may begin even before the preparations we have examined so far, and also explore various possible prolongations of demonstrative performances.

PRE- AND POST-DEMONSTRATION WORK

Many public demonstrations, especially the most highly publicized, are preceded by a considerable amount of advance "communication." This more or less catch-all term refers in particular to the way the performance is brought about and announced and to the way statements about it are formulated. One of the earliest goals is generally to arouse curiosity about it, even to create an expectation. It is a question of attracting a particular audience and making its members as receptive as possible to the event on the day it takes place.

We have already observed this type of phenomenon with the Scientists on the March street protest. The march in Paris had been broadly announced and publicized during the period when researchers were coming across France on their bicycles, and even before that.

But the same thing happens for other forms of public demonstrations, including demos.

To launch the NeXT computer, as we have seen, Steve Jobs organized a major public demonstration in Davies Symphony Hall in San Francisco in October 1988. The point was to show off the computer and explain for the first time to the public at large what it could do. To draw a crowd of journalists, photographers, and videographers to the vast auditorium (4,500 seats), Jobs opted for a communications campaign designed to release very little information on the technology. This strategy was aimed at generating high expectations, by suggesting that the technology was so revolutionary that its secrets had to be rigorously protected. The strategy also had the advantage of limiting the risk of disappointing the public, because care was taken not to mention specific features that might not actually be ready on the day of presentation—for example, a color screen (Lampel 2001; Elmer-DeWitt 1988).

Thomas Edison himself proved quite skillful at stirring up curiosity and creating an expectation regarding the technologies he was developing. He proceeded step by step, in an organized way (Lampel 2001). To reveal how his electric lighting system worked, he invited journalist Marshall Fox to watch a private demonstration in his laboratory in Menlo Park in December 1879. The article Fox published aroused a great deal of curiosity among a broad public. Edison went on to organize several demonstrations for limited audiences, especially for manufacturers and experts; these too received wide media coverage, and generated interest in the technology among a growing number of actors. The series of performances culminated in a big public demonstration whose audience was expected to be impatient, enthusiastic, and eager to see for themselves what they had been reading and hearing about for some time.

The media campaign that preceded the public demonstration of the Pioneer Zephyr train in 1948 was another instance of gradual distillation of information intended to create suspense and make the general public eager to see with their own eyes how this new technology worked (Lampel 2001). The demonstration consisted in completing the longest nonstop trip in railroad history, breaking the speed record in the process. Models of the train had been shown to journalists even before the train itself was operational. When the demonstration took place, impatient crowds were waiting at the train's point of departure in Denver and at its destination in Chicago, while other crowds positioned along its path cheered as it went by.

The development of hot-air balloons in the eighteenth century put comparable phenomena into play.[9] The public demonstrations of Montgolfières in France in the 1780s were preceded by a great deal of publicity, and the expectations aroused on the day of each flight also helped stimulate public curiosity and eagerness. For the launching of a hot-air balloon required extensive preparation: whether in the Tuileries Gardens in Paris or in the Royal Garden in Bordeaux, for example, crowds waited expectantly for hours on end, strolling about, picnicking, and discussing the event. This spatiotemporal organization, cumulating in the realization of prodigious feats long considered impossible, along with sensations generated by hunger, was the source of powerful emotions, above all a certain euphoria.

These examples make it clear that certain public demonstrations must be grasped as the culminating point of a series of prior events and actions. Thus it seems important to analyze the work that can be done well before the performances take place, especially if we want to understand how they are perceived, and if we want to account for the fact that the spectators' usual practices of critical evaluation may be abandoned in favor of the enthusiastic welcoming of a new technology, or fervent adherence to a new project.

Moreover, such investigations can allow us to spell out the effects sought by demonstrators, effects that are not limited a priori to arousing public enthusiasm for a technology. The goals of a demonstration can be multiple and subtle. For example, the announcements made before certain demos for the Delta project did not aim simply at arousing audience interest in the software. They also sought to avoid disappointing the public, for instance by refraining from making any promises that could not be kept. An electronic message drafted by a member of the Delta team and addressed to his colleagues illustrates the search for balance in the communication efforts made before the group released a demo. The researcher was trying to determine how to present the state of advancement of the project and the features of the program to a potential client whose understanding of Delta might be out of sync with the reality:

"Some of the . . . comments seemed to show a lack of understanding that we are very much in an experimental mode. . . . I think it would be best to send a message to [them] that reduced short term expectations, without deflating what appears to be substantial interest. Any ideas?"

This example stresses the importance that demonstrators could attribute to the stages preceding the demonstration itself. But it also shows that the ways audiences can be prepared for demos are not limited to mass communications prior to the launching event. In the case of projects implying more limited audiences, the work of communication can entail more sharply defined and personalized actions.

Moreover, we should note that there is no reason to suppose that the effects sought are always achieved, or that they are always received by the public in a homogeneous way. As we have seen, public reactions to demonstrations can prove to be quite different from those sought by the demonstrators; this is also true for preliminary communications. For example, Microsoft's 2015 TechDays had been announced enthusiastically in the press, as can be seen from this excerpt:

> Between February 10 and 12, 20,000 visitors are expected at the Palais des Congrès in Paris for TechDays, one of the biggest technology events in Europe. This 9th edition, under the heading "ambient intelligence," will also be followed by 100,000 Internet users. . . . Announcements about Microsoft's latest innovations can also be expected during this meeting. The realms of automatic translation and artificial intelligence will be at the heart of the event, always in the interest of promoting a more technological environment that will revolutionize our professional and personal lives.[10]

Following the advertising campaigns that preceded the TechDays event, I was able to observe a broad diversity of attitudes on the part of spectators before the public demonstrations began in a plenary session. The spectators did not all display great impatience or unbridled enthusiasm at the outset. On the contrary, their attitudes appeared varied and more or less moderate.[11]

Now that we have a clearer sense of why it is useful to analyze the actions taken around public demonstrations well ahead of the actual performances, we are ready to explore what could be called the postdemonstration phase. For grasping demonstrations and their effects cannot be limited to a temporal sequence reduced to the duration of the performance itself; we are a priori obliged to analyze them over a longer period of time. Demonstrations may well lead to a series of related actions in their aftermath, and even to other demonstrations, in the same or different forms.

To prolong the demonstrative actions carried out in the context of TechDays 2015, Microsoft encouraged the publication of a number of reports on the internet. The following excerpt provides a flattering picture of the technologies presented at the conference:

> If you follow me on Twitter, Facebook, or Instagram, then you probably know that Microsoft invited me to cover the event. I was unable to remain on site for all three days, unfortunately, but I confess that I was literally left breathless by this event. I won't lie to you, I was very afraid that I'd find myself in an ultra-corporate environment focused solely on the company's own technology. It may surprise you, but that wasn't at all the case. Of course it was impossible to take two steps without coming across a Surface, a Lumia, or even a PC powered by Windows 10, but what really stood out were some other companies' products. . . . But the real star, at least from my viewpoint, was obviously Windows 10. You can tell that this product has special importance in Microsoft's new strategy, and you have to admit that the company has set the bar very high. I actually had the chance to try the thing out on various devices, and I confess that I was pleasantly surprised by what it can do. Next year, for sure, I'll arrange my schedule so that I can stay all three days and cover the event as it deserves to be covered.[12]

This enthusiastic report makes it clear that Microsoft's demonstration of strength was not limited to the space of the TechDays event through the combined displays of its own products and those of its partners. It is also embodied in testimonials about the event that were disseminated through the media.

Other cases of actions that follow the performance of public demonstrations include the results of a study focused on the way a mobile telephone company used PowerPoint demonstrations. These were commonly sent out within the organization after meetings, serving as memos summarizing the meetings' contents and as tools for arranging follow-ups. They could be transformed by third parties to serve as the basis for new written or oral exchanges. Thus a work of interpretation and appropriation went on after performances were concluded; the performances in turn constituted preludes to further actions.

Similarly, at the 2014 Paris Fair I was able to observe how demonstrations of products were followed by specific efforts toward concluding sales (price reductions, bonus offers, and so on). At certain booths, the products purchased by clients were placed in transparent plastic bags; these products were thus visible as the buyers moved through the room. "Exhibits" like these were additional demonstrations of the interest that the objects in question had aroused. They were designed to convince other visitors of the desirability of such a purchase. In other words, the demonstrations carried out in the booths led to additional demonstrations.

The Scientists on the March event offers another illustration of the way a public demonstration can be part of a chain of many actions. As we have seen, the street protest in Paris was the result of other actions carried out in many places. The photos, videos, and media accounts of the demonstration themselves served to nourish the website devoted to the event, and they were useful elements in the organization of future demonstrations by the Scientists on the March association. Moreover, the street protest offered its spokespersons a legitimate platform for stating demands in the presence of the media, demands directed to political authorities, thereby producing new demonstrations.

In a more general way, a "successful" street demonstration, especially one that mobilizes large crowds, can supply a certain number of trump cards to the spokespersons of a movement or a union, so that they can remain or become essential interlocutors in dealings with the government or the press, thus pursuing a demonstrative activity on several levels (Champagne 1984). Moreover, the analyses made by political authorities and journalists of the nature, messages, and representativeness of a street protest represent an important series of actions that build on the event itself. This work of interpretation typically has no less impact on the fate of the movement than the protest itself.

This case reinforces my contention that the study of this type of practice benefits from being extended beyond the mere performance of demonstrations. Enlarging the framework brings in numerous elements that give us a better grasp of the practice. We have seen that the preparatory phase is often essential, especially from the standpoint of making the actual unfolding of the performances predictable. "Preparing" does not necessarily exclude improvising, however. In the case of public demonstrations constructed during interactions, it is often a matter of using easily mobilizable repertoires that can be brought in to complete the effects of elementary modes of socialization in the structuring of exchanges. Moreover, we have seen how the organization of a demo can entail a substantial number of operations, an iterative approach, and the involvement of groups that may be quite large. This activity can bring into play a significant division of labor and may fall under the umbrella of project management.

The projects in question sometimes stem from an undertaking of audiovisual production or from the construction of a highly publicized event of broad scope. In such cases it is easy to observe stagings, that is, the arrangement of spaces in which actors are to intervene in a coordinated and anticipated way so as to be observed by others. We have seen, moreover, that certain producers of public demonstrations use theatrical techniques, whether by design or otherwise: they mobilize various accessories, costumes, lighting designs, and other elements of the décor, and they organize various collective actions on multiple stages. Nevertheless, such a process is not universal. Some individuals do not associate the work of demonstration with an activity of representation or the organization of spectacles; some even actively resist such an association.

In any case, a given demonstration can entail any number of preparatory actions, including very careful management of announcements and of public expectations. In addition, we have seen that demonstrations are sometimes reused for new purposes or are followed by various types of interventions. These may take the form of a posteriori interpretations of what has been demonstrated, or of new demonstrations produced in different forms, designed to reinforce the earlier ones.

This last point warrants detailed exploration. Our next task is to examine an important dimension of these performances, that is, their possible inscription in a series of more or less orchestrated demonstrative actions and in demonstration arsenals. Not all public demonstrations are isolated acts. Their meaning and their effectiveness may depend, on the contrary, on their contribution to many sorts of demonstrative campaigns, as we shall soon see.

4

CAMPAIGNS AND ARSENALS

Many social phenomena spread out over time and space are commonly perceived as processes of short duration that occur in specific places.[1] Public demonstrations are no exception. A great many studies approach them as isolated acts and analyze them by focusing on very short spatio-temporal sequences. However, as I have suggested, there are many reasons to go beyond that framework. The actions that precede or prolong demonstrations can in fact help confer on them their full meaning.

Our task now will be to study the more specifically demonstrative actions that are apt to accompany public demonstrations. I shall examine the extent to which these latter may be inscribed in demonstrative campaigns made up of simultaneous or successive demonstrations and conceived to be deployed in concert. I shall focus on a variety of cases to show how sets of demonstrations can be carried out by several demonstrators acting in a coordinated way, mobilizing demonstrative arsenals that consist in multiple forms of demonstration selected according to the context. Let us begin, then, by examining how demonstrative campaigns may be organized.

DEMONSTRATIVE CAMPAIGNS

The development of the Delta project illustrates one way a demonstrative campaign can be carried out by a group of demonstrators. As we have seen, Delta sought to create software that would help prepare space missions. The project included researchers and engineers working for various research institutions in the United States.

To finance this project and to establish the basis for prolonging it in other forms, the members of the teams involved had to approach many different organizations. For these team members, Delta demos were an important resource as they sought to introduce themselves to possible funding sources. Proposing to present a demo was a useful tool for getting appointments with potential sponsors and, subsequently, for trying to persuade them that the project was worth of support.

A demo thus constituted, first of all, a mode of self-presentation. The first contacts among some of the participants in the Delta project took place during interactions of this sort. Two excerpts from messages exchanged by participants show how a demo could be substituted for other forms of encounter, self-presentation, and presentation of one's work, enriching the range of possibilities:

Sorry you didn't make the demo. [The demonstrator] isn't a great explainer, but it was pretty flashy, [the one] with the big screen. . . . Maybe there's some time we could get together? I could go over there or you could come for coffee here.

Sorry we couldn't make [the demonstrator's] talk. . . . We would like to have you visit us sometime over here. . . . Would you be interested in giving a talk here?

Even if it did not produce the same effects as meeting in an office, for example, or presenting a paper in a seminar, we can see here how giving a demo constituted an opportunity to present oneself and one's work. For the researchers, demos were an important tool for meeting with other actors in view of establishing relations of exchange. Moreover, each of these demos constituted a complete transaction: in exchange for having been granted an appointment, the demonstrators gratified their hosts with a display featuring the proposed project.

Creating a demo was thus not only an end point or an important step for the demonstrators in their theoretical and technological research (especially upon the completion of a version of their software), but also a means for discovering and approaching institutions that might bring them new resources. It was not simply a matter of validating an approach by exhibiting a working prototype, it was also a matter of using that tool to make contact with others and arousing interest in view of embarking on a partnership or gaining funding. These dynamics were just as much at work in the early exchanges among teams as during the teams' approaches to other institutions at a later stage.

The Delta project participants had thus embarked on a major demonstrative campaign. They had crisscrossed the paths connecting various businesses, institutions, and conferences. They had exhibited the workings of their prototype in various research centers and also in university seminars and colloquia.

This exploratory work was carried out in a systematic perspective. The idea was to cover a certain number of spaces in order to generate as much information exchange on the project as possible. It was not unusual for investment decisions to be made on the basis of indirect representations of a working prototype. The viewers of the demos were in fact witnesses in a position to share their appreciation of the project with broader circles than those actually present at the demo presentation sites. The demonstrators were well aware of that dynamic, as we can see from an excerpt of a message sent by a Delta project participant to a colleague: "I'll likely be seeing [him] at [the conference], we're tentatively planning on taking a machine to give demos. He could probably give a report to his colleagues, thus there might not be as much urgency to give a demo [elsewhere] right away."

This message illustrates the way the demonstrators managed the economy of their performances by taking into account the probability that witnesses would share information. It emphasizes the degree to which the Delta participants were concerned with multiplying their interventions. If the direct effects of these interventions were amplified by their number, it was also because the spaces approached were not watertight. Information circulated. A demo carried out in a given place, for example, in a lecture hall, could produce effects in other spaces. In other words, the demonstrative effort was capitalizable. The project's participants could manage their demonstrative investment in stages, in order to cover as vast and relevant a territory as possible, and in order to benefit from the cumulative interest of their various interventions, as long as these worked as expected.

In the context of this activity, demos were generally supported by other actions and forms of demonstration. These might entail more traditional communications giving rise to hybrid presentations, halfway between demos and interventions in a seminar. They could also entail statements exchanged in an office or a hallway among various individuals in positions of responsibility. The statements made in these exchanges could be collected in brief arguments prepared in advance, constituting a reserve of material for demonstrators to draw on. One-page summaries or brief formulations of the major thrust of the project were also distributed and sometimes posted on the internet. Alongside these research digests, teasers designed to arouse interest in the project on the part of decision-makers under time pressure, more extensive research reports were drawn up and communicated within organizations.

This demonstrative campaign was not developed by isolated individuals. On the contrary, it resulted from coordinated actions on the part of a group of demonstrators. Members of the group generally gave their demos in separate spaces in a concerted fashion. Nevertheless, they sometimes came together to present complementary aspects of their project. This formula made it possible to reinforce the spectacular character of the demos and to respond to a whole range of questions; no single member of the team had mastered all aspects of the project. Not only the quality and experience of the participants but also their sheer numbers could help impress the audience, and the accumulated expertise reinforced the credibility of the project. The explorations of institutions thus proceeded from demonstrations of strength.

But another key aspect of this sequential linking of public demonstrations lay in the participants' gradual production and collection of data, which they could use in various ways. Let us recall that the audience members at a demo were sometimes invited to manipulate the software themselves at the end of an initial presentation. On these occasions, the spectators who had become actors defined and dealt with cases that concerned them directly. The corresponding data were collected by the demonstrators. An excerpt from a message by a participant in the Delta project to one of his colleagues illustrates this process: "What we've been doing is to first run through one of the existing [examples] in single step mode, then we show them how to create one with the menus . . . and then we ask them to try entering their own [examples]. Thus we now have about a half dozen [examples] from end users in our library. . . . Unfortunately, the most complex [example] wasn't saved. . . . A simpler example of his was saved."

This passage makes it clear that demos did not consist solely in displaying what the software could do, but also in collecting cases formulated by the demonstratees. This undertaking allowed demonstrators to connect their targeted users to the future of the project, by showing them that the software was capable of accomplishing tasks that interested them personally. But this was not the procedure's only advantage.

For one thing, it gave the demonstrators access to examples that could be used in future demos, allowing them to show more clearly how the prototype could deal with "real" cases, how it could constitute an "operational" application in specific domains and be used for "concrete" problem-solving after a limited period of training. The activity of developing "good" demos in fact consisted in a full-scale research effort based on an iterative undertaking and on a systematic analysis of the unfolding of previous demos.

The data collected during demos (including during oral evaluations of the software and other verbal exchanges) also constituted useful support material for orienting research, "improving" or adjusting the software in response to the criticisms and interests expressed, thus increasing its chances for adoption. This is why this information was sometimes mentioned in demonstrators' reports, as we see in the following excerpt from an email exchange: "[He] and I are writing up a report summarizing the information we gleaned from the demos this past month. . . . It's not clear it will be done before I leave again on Saturday."

The data collected on the modes of appropriation of the software by the demonstratees were thus capitalized by the demonstrators in the context of the software development. This phenomenon appears even more clearly in the following excerpts from a series of messages in which a team member was sharing with colleagues his observations on the behavior of test users, the conclusions he drew for the evolution of the software, and further observations and experiments that could usefully be pursued:

The next version will incorporate what we learned from evaluations with potential end-users, as well as extensions we have planned all along.

What is the difficulty in explaining this? . . . When I've given talks and demos, it seems to enhance the audience's understanding.

Moreover, it only seems to take someone who is a domain expert about a half-hour tutorial to use the system themselves.

My experience with end-users reveals 2 kinds of . . . error: "deep" . . . errors . . . , and "shallow" . . . errors. . , , Diagrams wouldn't help very much on the former. . . . This is not to say that something interesting couldn't be done . . . but ensuring their utility would require more experimentation with end-users.

Analysis of the demonstratees' reactions oriented the research work toward making the software easier to use. The use of phrases such as "experimentation with end-users" suggests that, as in other innovative projects, the users could be tested as much as the technology (Pinch 1993, 35–37). For the demonstrators, the users' reactions during demos represented experimental material in its own right. These reactions, sometimes unexpected, were analyzed systematically. Thus the demonstrators paid attention to the time it took the demonstratees to master the software and to the ways they understood how the software works, as the excerpts cited show.

This multiform capitalization marking the demonstrative campaign of the Delta project participants could also be observed in one of its later phases. After several years, some team members had the idea of developing a version of the software for pedagogical purposes. This version was designed to explain the dynamics of the planets in the solar system and to support various educational activities in the classroom, among other goals. Demos were put online. The intended audience consisted primarily in students and teachers in American schools, from the elementary through the secondary level.

Several pedagogical scenarios and animations were proposed. The teachers and students who saw the demos were invited to share their reactions and suggestions, to help in the development of new scripts. The exercise undertaken by the demonstrators, then, was of the same nature as the one they had done earlier. The point was to collect information on the practices of their future "partners" or "clients" and to glean suggestions that would help define new cases and that could be capitalized in the form of new demos and future versions of the software. It was also a matter of moving ahead with both a program and practices (in this case pedagogical practices) and of constituting a relational web with promoters of the software, or even of creating a market for an innovative product.

The demonstrative campaign carried out by the members of the Delta project and the capitalist endeavors that characterized it were thus deployed over a longer time span and had a more proteiform aspect than one might have supposed at first glance. At the price of continuous adaptations, the actors drew benefits from their demonstrations over several years. But these researcher-entrepreneurs were not the only ones to realize capitalization—especially by accumulating various forms of credit—over the long run from their demonstrative campaigns. Their institutions also benefited, as did the leaders charged with accounting for their budget and for legitimizing their management, to the extent that these factors helped ensure their own promotion.

In the case of the Delta project, the demos thus did not constitute isolated demonstrational moves made by unrelated individuals. They were achieved in waves by an organized collective to cover a series of spaces deemed pertinent and to draw various benefits from them. To be able to appreciate their conditions of possibility and their stakes, it is important to grasp them as a whole and as skillfully orchestrated actions.[2]

To enrich the analysis further, let us look at an example of a demonstrative campaign carried out in a different context, that of the debates over a theorem that we have already considered (Rosental 2008). Let us recall that this theorem called into question the foundation of fuzzy logic, an important domain in the field of artificial intelligence. A large number of actors had gotten involved in discussing its validity and had proposed new demonstrations or counterdemonstrations, many of which were posted on a dedicated electronic forum.

Certain interventions, as we have seen, benefited from greater visibility than others, owing above all to citations and frequent repostings. In fact, the actors proposing demonstrations and counterdemonstrations were often not isolated participants: many of them intervened in the context of collective actions, implying a significant work of demonstrative coordination.

Two large groups of actors were in confrontation, the partisans of fuzzy logic on one side, and researchers favoring other approaches to artificial intelligence, as the original author Charles Elkan did, on the other.[3] In this context, most of the major figures in fuzzy logic had come together to offer as effective and visible a response as possible to an article that seemed to threaten the credibility of their field of research. They sought to avoid producing redundant—or, worse still, contradictory—objections, as had happened in the reactions initially posted on the forum by researchers in fuzzy logic. Thus some major figures in fuzzy logic published more unified counterdemonstrations and strove to make them highly visible and accessible. Their undertaking was extended in specialized journals, where a federation of counterdemonstrators, in articles attributed to as many as eight coauthors, responded to Elkan's theorem.

Thus the demonstrations we are analyzing did not stem simply from occasional actions by isolated individuals but rather from orchestrated actions reflecting group dynamics. A demonstrative campaign had been set up involving a federation of actors intervening in a coordinated fashion in various arenas, either simultaneously or in succession. The public demonstrations at stake took a relatively homogeneous form, since they usually featured proofs formulated in writing and published in forums and journals. However, each public demonstration had to be interpreted in relation to the overall architecture within which it was situated, and in relation to conflicts of diverse natures that were taking place over time on a broad scale and that entailed multiple actions.

Many such demonstrative campaigns are available for analysis, including cases involving product launching, such as that of SunSoft's NEO system, as we have seen. The development of the NEO demo by a dedicated team was in fact the prelude to a major demonstrative campaign, which was based above all on a multiplication of demos in numerous spaces and involved a large number of demonstrators.[4]

Similarly, during Microsoft's TechDays in 2015, the multiplication of demos during lectures and in the booths in the exhibit space could be understood as a demonstrative campaign on the part of the company that organized this multiday event. The demonstrations were produced by Microsoft employees or by representatives of the partner companies meeting at the conference. While the individual or collective demonstrations (some were carried out by groups of demonstrators, especially during the plenary sessions) served the interests of each demonstrator, they also had cumulative effects. The simultaneous performances served to draw a large number of visitors into the same space. Taken all together, they demonstrated Microsoft's strength in technological innovation; this was emphasized, for instance, in the commentaries produced in the context of the 3D printer demonstration that we examined earlier.

We can also look at cases of street protests: far from constituting occasional, isolated acts, these are commonly inscribed in series of actions that mobilize groups of variable size, working over an extended period of time. The groups involved may organize campaigns involving several forms of public demonstration, including but not limited to additional street protests. As we have seen, the organizers of the Scientists on the March event undertook related demonstrative operations over time, including further street protests and other public demonstrations on their website.

Thus we are beginning to see, via these cases and the previous ones, how demonstrators can call upon what can be called arsenals constituted by several types of demonstrations. In the discussion of specific examples that follows, I use the term "arsenal" first of all to characterize uses of diverse forms of demonstration in situations where competition is involved (competition for funding, for imposing one point of view over others, and so on). However, since such uses may be observed in other situations, I also use the term by extension to characterize other grouped forms of demonstration mobilized for various purposes. Let us now look at some arsenals in more detail.

DEMONSTRATIVE ARSENALS

While studying the work of researchers in artificial intelligence at Stanford, I observed that someone presenting a thesis in computer logic in the 1990s typically prepared a written demonstration of the properties of a new logic, used this logic as the basis for a computer program (a program for demonstrating theorems, for example), and then put together a demo of the corresponding software. This demo might be addressed to peers, when it was developed in the academic context of defending a thesis; it could also be addressed to the industrial sponsors who had financed the student's work (Rosental 2007). The demonstrations produced by the young researchers at Stanford (including their demos, demonstrations of logical results intended for publication) thus had to be grasped in the framework of the demonstrative arsenals in which they were situated. The elements forming these arsenals were mobilized selectively, and they could be combined according to the needs of the moment.

Architects constitute another category of professionals endowed with rich demonstrative arsenals. Their demonstrations can be based on mock-ups and plans, on computer simulations, PowerPoint presentations, or displays of sample materials, among other things. But these demonstrative arsenals hardly outstrip those of protest movements in scope: the latter can incorporate not only peaceful street protests (as in the case of Scientists on the March), but also, and sometimes simultaneously, the creation of barriers, the occupation of particular sites, vandalism, or even armed confrontations (Fillieule and Jobard 1998). The demonstrations of strength brought into play can depend just as much upon the mediatization of images from marches as upon figures indicating the high number of participants in the event (Champagne 1984).

The demonstrations produced by national space agencies for the public at large offer yet another example of particularly large demonstrative arsenals (Rosental 2007). Widely disseminated photographs and videos of the solar system, along with assorted arguments in favor of pursuing space programs, are complemented by astronauts' demos showing details of their daily lives in space. Compared to other forms of demonstration, the astronauts' demos seem particularly effective in arousing public interest in the conquest of space, as we can see from those produced by Chris Hadfield during his sojourn on the international space station from December 2012 to May 2013. The first Canadian to head this station, Hadfield posted tweets and videos on YouTube showing how he went about his daily activities. Responding directly to questions asked on the internet, he had a public of a million Twitter followers. A newspaper account stresses Hadfield's intuitive gift for choosing the most effective forms of demonstrations (cited in Sutcliffe 2013):

Hadfield executed a flawless, sophisticated marketing campaign. He did this for the space program without ever talking about the cost and benefits of exploration. This wasn't an advocacy campaign where he droned on about the value of science and research. It was a masterful content-driven strategy of spectacular photos, fascinating observations and clever videos that made life on the space station relevant to people who had never been there. He singlehandedly rejuvenated interest in space. Hadfield understands that the way to convince people to believe in something is not to present a rational argument, but to get them to care about it. . . . He played to his audience by tweeting pictures of cities as he flew over them and connecting to what was happening on Earth as much as what was happening on the space station. . . . His tweets were tight and clever. His videos were visual, brief and amusing. He didn't try to do too much with any one message. How many times have you seen an ad or a billboard cramming in so much content that it's unreadable? . . . The beneficiaries of Hadfield's campaign may be space exploration, the International Space Station or the Canadian Space Agency. But the product is him. . . . What brand wouldn't want to be associated with Hadfield's next mission?

This article emphasizes the importance of the choice of forms of demonstration in one particular case, and the relative pertinence of one form or another depending on the context. Compared to quantified demonstrations or lengthy arguments formulated by experts, demos of the daily life of an astronaut in space, accompanied by pictures and succinct commentaries, seem particularly effective and adapted for promoting the space program—as well as the demonstrator himself—with the public at large as the targeted audience. Nevertheless, well-developed and quantified demonstrations remain relevant for certain audiences, especially administrative and political authorities (Rosental 2015). Having well-stocked demonstrative arsenals therefore is an asset for space agencies such as NASA (Jouvenet, Lamy, and Saint-Martin 2015); these arsenals allow the agencies to call up forms of demonstration that are adapted to the circumstances.

This example raises more general questions. On what basis can a given form of demonstration or a given demonstrative arsenal be privileged by individuals or groups in a given context? What constraints weigh on these choices, and what is at stake? Thus to the question of the effects of the combined uses of diverse forms of demonstration we must add the question of the extent to which demonstrators have a sense of what demonstrations they should use, and what accounts for the selective recourse to certain forms of demonstrations for certain audiences. To approach these questions, it will be useful to return to the demonstrative campaigns carried out in the context of the Delta project.

We have already seen how the team responsible for developing the Delta software used a variety of forms in its demonstrative campaigns. Its public demonstrations varied in size, from summaries designed for websites to more elaborate arguments presented in research reports. Demonstrations of the Delta project could take written or oral forms, and they could mobilize sophisticated equipment.

Each format was adapted to a specific situation. For example, brief oral arguments were often stabilized and used as resources by researchers interacting with managers in informal settings. Arguments written down in a few lines were better adapted to texts presenting the project in advance of a demo or following a discussion.

The oral exchanges in which project managers could interact with demonstrators, or with third parties whose opinions were being solicited, called for demonstrations that were more substantial although still limited in duration. Demos filled this role perfectly. Thus arguments of just a few sentences relating to the Delta project, whether presented over the phone, in an office, or in a hallway, were good preludes for demos. They permitted an initial communication about the content of the research. Brief exchanges also offered demonstrators the possibility of refining the formulation of responses to calls for bids designed to fund the project.

As for the research reports, they allowed evaluators to consult particular details as necessary. Finally, the written summaries sent out before or after a demo could be reproduced or adapted by team members in the process of formulating responses to calls for bids. The following excerpts from messages exchanged by certain team members provide good illustrations of these proceedings:

I'm giving a demo to [a manager] this afternoon. We might also possibly discuss general aspects of [next fiscal year's] funding for outside efforts. . . . It would be good to think of specific topics that might be proposed. While I could use your portion of [the] proposal as a starting point, if you have other topics more closely related to our collaboration that you would like to pursue, perhaps you could forward me brief descriptions (i.e. about one paragraph per separate topic, no more than three topics).

It would be good to at least have a 1–2 paragraph abstract in hand . . . that could be quickly expanded into a 2–3 page preproposal if extra money materializes. Let's aim for Monday . . . for a preproposal abstract. If you have time, a full preproposal would be marginally better.

I gave a talk to the . . . group about our work, and also described [the] grant. I talked briefly with [a manager] in charge of your grant.

These passages highlight the breadth of the demonstrative arsenal on which the Delta project participants could draw. They show how the participants were led to shore up demos by informal discussions with managers in order to obtain funding. This practice was also supported by the distribution of succinct summaries of the project. The messages also brought to light the combined use of traditional means of communication and informal exchanges.

The Delta project team members used a large array of research teasers, all kinds of short presentations designed to arouse the interest of busy decision-makers, summarizing the project's main outlines and salient features. These presentations were developed and put to use adeptly, over time. Their effects were amplified when they were combined with one another. For example, following a spectacular demonstration likely to stir up enthusiasm in a decision-maker, a discussion in a hallway could help dissipate doubts or stimulate debate over the practical modalities of attributing funding. Distribution of a summary then offered an opportunity to anchor a lasting favorable impression in a manager, and to supply the latter with the supporting material necessary for a bureaucratic activity, based in particular on the creation and circulation of files.[5]

Taken all together, the demonstrations in play appeared partly interdependent and could thus be all the more readily grasped as constituting an arsenal, in that their constitutive materials were transformed by team members in order to be used in different formats. Each format was in fact adapted not only to a particular situation but also to particular interlocutors. The ingredients of the demonstrations were used more than once. They were selected, adjusted, reshaped, then reassembled. Fragments of written presentations were thus mobilized to form oral arguments and vice versa.

This production was fed by active trading of summaries featuring the research. Elements of PowerPoint presentations and summaries were exchanged, as we can see from the following excerpts from correspondence among team members:

I'll forward a talk abstract I used . . . feel free to cut and paste.

By the way, if you have one or two slides about [this] in postscript format, I might be able to work them into my talks next week.

In my talks I haven't discussed the details of [this]. If some of your slides deal with that, it might be worthwhile seeing them to see if I could incorporate them in my talk. (I'm already showing [three people's] slides).

These excerpts make the traffic in slides among demonstrators quite obvious. The demonstrative arsenals were ultimately constituted via such repetitions, transformations, and recombinations. They resulted from capitalization of the demonstrative efforts, and they were linked to an economy of reuse that was being deployed on a broad scale. This economy also included demonstrators' collections of problems formulated by spectators during or following demos, as we have seen. These problems offered demonstrators new examples on the basis of which new demos could be developed. The recuperation of supporting materials for demonstrations went beyond the framework of the Delta project, moreover; it was practiced on a broader scale.

This economy of recuperation was clearly valuable from the marketing standpoint, at least as Carmine Gallo saw it, for instance (Gallo 2010). It allowed the same messages to be hammered in with the help of varied communication tools, so consistent information could be circulated in a variety of spaces and in a way that would leave its mark on recipients' minds.

Practices of this sort invite us to consider once again the capacity of a demo to disrupt the critical habits of its viewers, extending the question beyond the context of a given performance into a longer time frame. If demonstratees feel enthusiastic about a project when they are attending a demo, what happens when they are later exposed to other forms of demonstration in the context of the same project? The effects of a demo, and in particular the conviction that it is capable of generating, may be temporary. These effects may be reinforced or, on the contrary, countered over time when demonstratees witness other forms of demonstration.

It is appropriate, then, to be prudent about generalizations according to which demos make it possible to win over their audience members by transforming them into irrational beings, or even into mindless enthusiasts incapable of exercising any critical faculties. On the contrary, judgments have to be made on a case-by-case basis; one must not generate artefacts linked to a defective methodology. Instead of grasping demos solely at the point of performance as autonomous sequences and isolated acts, we need to analyze them a priori as elements of potential combinations whose effects must be measured over an expanse of time and space. This appears all the more necessary in that there is no reason to suppose in general that decisions to invest in a technological project are made on the basis of a single demo.

The demonstrative arsenal deployed in the Delta project can be compared usefully to the one put into play in a European research program called Advanced Communication Technologies and Services (ACTS).[6] I was able to observe and analyze the implementation of this program, which was located squarely in the field of information technology. It brought together various types of participants from a number of European countries: researchers, engineers, and executives, many of whom were working for telecommunications and computer companies (Rosental 2013, 2017). The explicit principal objective of ACTS was to develop a high-speed communications network in Europe, along with adapted multimedia services, in order to contribute to economic development in Europe and to increase the competitive position of European firms. More specifically, ACTS was intended to support the development of telework within large companies and within European institutions, by making it possible to hold high-definition video conferences. For this purpose, with a budget of about 700 million euros (roughly a billion dollars), the program selected and financed more than 150 projects.

As these projects advanced, many public demonstrations were produced in order to "disseminate information," to borrow the expression used by the ACTS administration. These demonstrations were intended for the masses of European citizens whose ability to understand science and technology was assumed to be limited, but also and to an even greater extent for journalists and representatives of political and economic authorities. They were developed within a specific political context. The ACTS leaders were confronted with insistent questions, competing demands, and criticism related to their budget management, coming from members of the European Parliament and from various industrial lobbies. These latter, like the lobbyists for the telecommunications industry, were trying to obtain the maximum possible level of financial support from the European Commission.

Under these conditions, the program managers had to show that their arbitration was fair and appropriate. Highlighting the productivity of each ACTS project was also an essential card to play. These managers thus asked the program participants to demonstrate the feasibility of their projects and to show that they were bringing in "concrete" and "useful" results. For the ACTS management team, this was a way of showing that the funding allocated to research and development was indeed producing results of economic value. This undertaking informed the expression adopted to characterize the program: "Programme for Research and Technology Development, including Demonstration." This formula gradually came to cover the entire set of European research and development programs.

Under this expansive label, the term "demonstration" was used to cover several forms of public demonstration. Some of them were presented in many technical reports and scientific publications produced by ACTS program participants. They also appeared in summary reports presenting the results of the program, supported by various statistical charts and graphs. These tools allowed relatively compact and readable documents to circulate among political authorities, industry leaders, and members of the European administration.

In order to demonstrate the relevance and tangible character of the ACTS results, these reports offered illustrated summaries of the projects. They contained lists and statistics related to the publications, patents, technical standards, and experiments produced in the context of the program. They also presented the results of surveys carried out through questionnaires addressed to project leaders, spelling out the goals the latter had indicated and the benefits they thought they had gained from their participation in ACTS.

The production of statistics had various advantages for the program leaders. It showed that these leaders were evaluating the project results regularly in a professional and transparent manner. The figures were the external signs of a serious, systematic, and objective examination of the progress of the research under way.[7] Moreover, they constituted a precious resource to counter the criticisms that could be formulated about

ACTS management and financial arbitrations. In case of need, European Commission officers could call up a certain number of figures in the context of exchanges and public responses and could refer to the reports that brought them together. In other words, the statistics were available and the ACTS leaders could use them to try to demonstrate the legitimacy of their actions.

These figures, like the reports in which they were included, nevertheless had certain limits. They had to be read, whereas the intended audiences often appeared to prefer images to texts and interaction to solitary reading. Prose texts, like statistics, were perceived as relatively unappealing, hardly capable of capturing attention and arousing enthusiasm. The production of figures and reports were thus completed by other actions.

One of these consisted in constructing and periodically updating electronic databases on the content of the funded projects, so the information would be accessible on the internet. Summaries expressed in relatively unspecialized language were preferred, along with images.

The program leaders also organized the production of "success stories" that recounted in concise form the prodigious success of certain projects. These stories were distributed, for instance, in the form of brochures or CD-ROMs, under the title *ACTS Multimedia Success Stories*. Journalists deemed by European officials capable of attracting a broad audience and arousing a certain "fascination" with the program results had been asked to contribute to this production.

To complete this demonstrative arsenal, multiple demos were organized, crucial tools that could demonstrate to European political representatives the reality and relevance of the program results. The ACTS leaders had thus required the program's participants to produce a significant volume of demos. A number of these demos had been in the planning stages for a long time and were to be completed according to a specific calendar.

Certain demos amounted to showing off high-speed information exchanges that would allow diverse forms of telework. They brought together executives and leaders of telecommunications and computer technology companies, engineers, researchers, high officials from the European Commission, lobbyists, journalists, and politicians from various European countries.

One illustration of the way European officials were proceeding can be found in a teleconference organized in Brussels titled "Twenty-First Century: The Communications Age," which offered images of exceptional quality, intended in particular to demonstrate the successful working and the usefulness of the prototype high-speed network financed by the program.[8] Bringing together actors located at the European Parliament in Brussels, in various European countries, and also in Japan and Canada, the event highlighted the achievements of several projects funded by ACTS. Demos of multimedia projects were enhanced by the outstanding quality of the images displayed.

For ACTS managers, organizing this teleconference involving political and economic leaders from across the globe constituted a powerful, targeted tool for demonstrating the existence and the value of ACTS projects' results to actors attentive to public policies and expenditures but unable to evaluate material presented in specialized scientific and technical articles. These participant observers were thus not forced to form opinions as to the productivity of the ACTS program solely on the basis of expert opinion, the weight of accumulated publications, or the length of the lists of publications presented in progress reports. Their more or less moderate taste for reading and for statistics was not put to a test, either. The type of demonstration provided by the conference could respond instead to a more pronounced interest in animated images and personal interactions.

Nevertheless, not all the demos were performed "live." Some had been recorded in video format. These were made accessible on the internet or distributed to industrialists, political representatives, and journalists in CD-ROMs. The multiplication of demos in various formats was designed to increase the visibility of the program significantly. In order to broaden the circles of spectators, accounts of these presentations were published, most notably as program progress reports. The circulation of printed accounts helped make the demos visible to actors who had not had the opportunity to see them in person.

Thus, for example, an ACTS progress report gave a detailed list of the public demonstrations carried out by the program participants. The demos of the various projects were mentioned sequentially in three-column charts indicating the date, nature, and target audience of the demonstration, along with notes on the audience reaction. Demos carried

Table 4.1 Excerpt from a Progress Report Listing ACTS Projects' Demos

Date	Nature of Demonstration	Target Audience and Reaction
9/02/96	Showing of *Eye to Eye* on dual display and sequential TV receiver to national press and technical journals.	Everybody most impressed with the demonstration. Result was a number of published articles.
9/12–16/96	Showing of *Eye to Eye*, demo videos of virtual production, live demo of Virtual Edit Suite (VES) and character animation.	Most considered this the best 3D TV demo they had seen. Much interest in the virtual studio systems and VES.
9/18–21/96	Showing of *Eye to Eye* on large-screen projection and sequential TV. Videos of MIRAGE virtual production and virtual characters.	Politicians, academics, broadcasters, and manufacturers were highly impressed by the standards achieved and the practical systems demonstrated.

Source: ACTS, *Results, Impact and Exploitation,* Interim Report, European Commission, DGXIII/B Ref: AC 1997/1339, May 15 1997, 40.

out in the context of a project called MIRAGE—Manipulation of Images in Real-time for the Creation of Artificially Generated Environments—are summarized in table 4.1, above.

This table highlights the importance and diversity of the audiences for ACTS demos, along with the breadth of media coverage. It makes clear the frequency of their performances—three times in a month, in the case examined—and the spectacular aspect that they might take on. It also shows how audience reactions were systematically described and analyzed, emphasizing in particular the viewers' enthusiasm.

Table 4.1 helps bring to light the vast demonstrative arsenal that the ACTS leaders were seeking to set up—and of which the table itself was a part. The recourse to this arsenal, with differentiated uses of diverse forms of demonstration according to the context, made it possible to produce a demonstration of strength on a very broad scale. The various demonstrations, as they accumulated, served to underline the innovative power of the program and its participants.

The ACTS case, then, is an example of recourse to a larger demonstrative arsenal than those we have studied up to now. This has to do with the scale of the program, which extended much further than that of a project bringing together just a handful of participants.

The use of demonstrative arsenals of similar magnitude can be observed in the context of other European programs and in major technological projects, especially in the energy sector. This is particularly the case for projects developed in recent years for the capture and underground storage of the carbon produced by industrial activity, in order to limit its dissemination into the atmosphere and thus to limit the warming of the planet. In cases like these, as in many other cases relating to the development of technologies that have major environmental stakes, projects frequently run up against opposition. The public demonstrations on one side, in the various forms, are thus often confronted by those from opposing sides, in equally varied forms. Demonstrative and counterdemonstrative arsenals then enter into competition.

COUNTERDEMONSTRATIVE ARSENALS

Recent carbon storage projects offer a first example of demonstrative arsenals in confrontation. These projects have been developed chiefly in the context of major demonstration programs financed by nation-states or groups of states such as the European community. They entail very high costs and high risks for industrialists. National and international programs are principally aimed at demonstrating the feasibility of technological approaches through the elaboration of what might be called, in a first approximation, prototypes.

There is no shortage of carbon storage projects. Since 2005 in particular, a number of them have been associated with the renovation of old coal-based facilities or with the development of new ones (Russell, Markusson, and Scott 2012). One example is the Logannet project in the United Kingdom, initiated with the goal of rehabilitating a coal-based power plant and transporting the carbon to a reservoir located under the North Sea (Markusson, Ishii, and Stephens 2011, 297). Another is the FutureGen project in the United States, intended to create a next-generation coal-fired power plant that would emit very little carbon thanks to a storage system on site (295–296).

These large-scale devices are often characterized as "demonstrators." They aim to demonstrate by their very existence and their performance that they bring a relevant technological solution to a particular problem: the dissemination of industrial carbon into the atmosphere.

Some attempts to store carbon underground have run into difficulties, ranging from leaks to explosions resulting in large numbers of casualties. In order to appear "satisfactory," a demonstration of an underground storage device therefore must show not only that its production costs are acceptable, but also that it is watertight and risk free over time (Markusson, Ishii, and Stephens 2011; Russell, Markusson, and Scott 2012). The audiences targeted by such demonstrations consist in political and administrative authorities, populations living near the proposed installations, journalists, members of nongovernmental organizations focused on environmental issues, and industry leaders in the sector (Markusson, Ishii, and Stephens 2011, 294).

To the extent that the capture, transportation, and storage of carbon entail the development of a great many technological devices, some of which are very large in size, the presentation of the latter may consist in numerous demonstrations in different spaces and in diverse forms. Unlike smaller-scale devices such as human-size robots, these large arrays cannot be presented via a demo in a room, in a face-to-face encounter with an audience. However, demos of selected components, simulations with the help of scale models or computer software, or discursive presentations illustrated by charts and graphs can be envisaged in cases like these.

In this respect, carbon storage technologies can be compared with those of the ACTS program, which were both diverse and, in some cases—such as physical networks for transporting data—very extensive. For projects like these, demonstrations have to be widely distributed in time and space and have to call on a large demonstrative arsenal (Russell, Markusson, and Scott 2012). For carbon storage technologies, demonstrations can take the form of visits to power plants, publications attesting to their performance, public lectures delivered by experts, and so on. Setting up an on-site information center, as has been done for FutureGen, also creates an important demonstrative resource.

Given the technological conglomerates brought together by these large-scale demonstrators, it proves quite complicated to produce consensual interpretations of what is being demonstrated and to generalize about the results. Moreover, given the risks that these technologies present, it is not surprising to encounter many opposing voices and counterdemonstrations on the part of local populations and environmental NGOs. As an example, let us look at the opposition to a European program that financed the development of several large-scale demonstrators, New Entrant Reserve 300, or NER 300 (O'Neill and Nadaï 2012). The counterdemonstrative arsenals consisted in arguments disseminated on the internet or through journalists, and also in protest actions around the demonstration sites by NGOs such as Greenpeace. These arsenals were as diversified, as widely distributed, and as extended in time and space as those of their adversaries. Their complexity made controlling the audience and its reactions particularly complex and uncertain (Russell, Markusson, and Scott 2012).

France's nuclear power program also illustrates the way in which demonstrative and counterdemonstrative arsenals can be constituted from an antagonistic perspective. The demonstrative arsenals of those who oppose the nuclear program have evolved significantly since the program was launched, shifting focus, depending on the period, from street protests to reports by experts to public announcements of radioactivity levels. Confronted by these diverse forms of demonstration, the public relations service of EDF—France's leading national electric power company, which manages a large group of nuclear plants—has developed a significant counterdemonstrative arsenal since the service was created in 1971.[9]

Thanks to an ample budget, its members engaged in an undertaking that mixed popular science with various communicative actions designed to promote nuclear energy. In particular, starting in 1972, EDF organized visits to nuclear power plants for local elected officials. A form of nuclear tourism was initiated in the mid-1970s, with more than 72,000 visitors in 1975; by the 1980s, this number had increased to some 300,000 people a year. The nuclear power industry has organized far more visits to its facilities than any other industry in France.

The public was thus invited to see for itself the absence of risk in these plants and to admire a major technological innovation. The visits combined several forms of demonstration. During a guided group visit at the nuclear plant in Saint Laurent des Eaux in which I participated as a schoolboy in the late 1970s, a demonstration of the workings of several parts of the facility was offered. It tended to make the plant look like a vast network of pipes, and the spaces being visited showed no signs of danger nearby. The fact that we could see and touch parts of the apparatus tended to make it seem familiar, comparable to any ordinary plumbing installation. The fact that we had to go through airlocks and wear disposable shoe coverings along with little instruments measuring radioactivity helped to demonstrate the extent to which the safety of these spaces was assured. The explanations provided by an easygoing guide about how the plant worked also contributed to the demonstration of the facility's harmlessness.

This demo-visit was preceded and followed by talks bearing on the overall workings of the plant and the parts that could not be visited, as well as on the general outlines of the nuclear sector. The oral demonstrations were coordinated with presentations of audiovisual documents. Given the time it took to usher school groups through the site, a full day was required for the outing. And the visit itself was extended by the distribution of various brochures that were later used as the basis for pedagogical activities in class.

Elementary and high school students and also university-level students in engineering were the privileged public for this type of visit in the 1970s, representing nearly half of all visitors; their numbers increased in later years. Journalists were another important group targeted by the EDF public relations service.

Nevertheless, EDF also developed other tools for its demonstrative arsenal during the 1970s. These latter took the form of written and audiovisual documents: brochures, slides, pedagogical films, and also documentaries, some of which were broadcast on television. This production was primarily designed for elected officials, doctors, teachers, and university students, but many of its elements were also accessible to the citizenry as a whole. EDF also organized talks on request for middle school and high school student groups.

The demonstrative arsenal deployed by the French power company thus was quite extensive.[10] Visits to power plants constituted the main pillar. These visits, at first sight, put an absence of danger on display.[11] They also put on display the presence of extreme security measures. Furthermore, they tended to make tangible, familiar, and almost ordinary spaces that discourse alone, or mere televised images, would have been hard put to make less abstract and less disturbing. Over and beyond their reassuring character, these visits helped generate public admiration for a sophisticated technological innovation, and to make it appear as a source of benefits for humanity.

This particular demonstrative arsenal has proved extremely effective. It seems to have constituted a powerful apparatus for mobilization, making a big contribution to winning support from the younger generations for the nuclear program. At the very least, it helped to defuse challenges for a long period of time. In the public opinion surveys that EDF carried out over the years, the company seems to have seen opposition to nuclear power fade away to such an extent that its very costly educational program no longer appeared necessary. In fact, the program had required the deployment of quite a large staff, as well as considerable logistical means, whether for organizing on-site visits or for arranging mass transportation to the site. The September 11, 2001 attacks in New York led EDF to put a stop to public visits to its power plants, at least temporarily.

That decision may help explain the surge of protest actions against nuclear power in France in recent years. By contrast, the American space agency NASA, which mobilized considerable means of communication to promote the space industry from the outset, taking specific steps aimed at journalists and creating educational programs to win the long-term support of generations of Americans, has never experienced such a shift. Its relations with the media and its partnerships with public schools have continued uninterrupted. Even if public opinion about NASA has altered somewhat over the years, especially after the explosion of the spaceship Challenger, opposition to the space program seems to have been on the whole quite restrained.

The effects of the demonstrative arsenal deployed by EDF over several decades must certainly not be underestimated. The same can be said about the costs of the demonstrations produced. This latter observation raises the question of the pertinence of Erving Goffman's assertion that demonstrations must put limited costs into play (Goffman 1974). In the cases I have just described, we have seen that the costs can be, on the contrary, quite high.

The EDF example brings to light the way in which struggles can rely on the constitution of demonstrative and counterdemonstrative arsenals on a broad scale. Energy is not the only industrial sector concerned, of course. The recent demonstrative battles that have accompanied the development of nanotechnologies in France offer another good illustration.

The Metropolitan Area Council of Grenoble, which is an important site of nanotechnology development in France, had charged the NGO Vivagora with organizing a series of citizens' debates over these technologies in 2006, with the participation of various experts, in order to respond to the opposition that had been expressed. According to a study of these processes (Laurent 2011, 2017), the public demonstrations produced in this framework ran up against opposition from the members of the PMO (Pièces et Main d'oeuvre [Parts and Labor]) association, which had refused to participate in the debates, denouncing what it deemed manipulations designed to favor the acceptance of nanotechnologies, the development of superficial products, and the legitimization of choices already made. PMO activists had instead organized a series of counterdemonstrations.

These activists drew up and publicized a chart of the connections between the relevant local authorities, industry leaders, and the scientists responsible for developing nanotechnology projects, to show how certain local decisions had been made by a small group of people. They "occupied" a crane at the construction site where a center for a large French research organization in the field of nuclear power was being built. They also published a fake newsletter purporting to emanate from the Grenoble Metropolitan Area Council, in order to criticize in a humorous way the measures of citizen controls that nanotechnologies would allow.

Their counterdemonstrative arsenal included still other resources. PMO members organized parallel debates aiming to turn passive spectators into active citizens critical of nanotechnologies. To that end, they constructed a public space composed of websites, independent media, and meeting places in the Grenoble area. In addition, they interrupted various scientific conferences focused on nanotechnologies and tagged walls in the city with slogans hostile to nanotechnology.

This example shows how conflicts can mobilize demonstrative and counterdemonstrative arsenals of different natures. The responses made to certain public demonstrations can imply counterdemonstrations that take distinct forms.

Such phenomena can be observed in many other situations, especially where environmental issues are concerned. Demonstrators can just as well be NGO members or scientists as industrialists, administrative or political authorities, or journalists. Each can take recourse to specific demonstrative arsenals to defend antagonistic interests, for example in the context of building industrial complexes or roadways. Public demonstrations can take the form of street protests, site occupations, public proofs of scientific facts, or even "happenings" (Rosental 2011). The mobilization of bodies, science, or art can serve to bring to light the existence or the absence of danger or the presence of various sorts of public problems.

The environment is of course not the only realm concerned by such political actions. As evidence, it suffices to consider the demonstrative confrontations spurred by Colin Powell's public demonstration, at the United Nations headquarters in February 2003, of the presence of weapons of mass destruction in Iraq, the purpose of the demonstration being to justify sending troops to that country. The PowerPoint presentation made on that occasion was a key element in Powell's demonstrative arsenal. Certain segments of that presentation, especially satellite images of sites located in Iraq that were presumed to house weapons of mass destruction, were reused to construct counterdemonstrations disseminated via the internet (Stark and Paravel 2008).

Academic exchanges are not exempt, either, from the mobilization of demonstrative and counterdemonstrative arsenals. For example, I have observed how researchers in artificial intelligence could invoke theoretical results resulting from demonstrations published in journals in order to challenge assertions that had been made in demos. Conversely, I have seen researchers invoke the successful performance of demos to cast doubt on the validity or verisimilitude of results demonstrated in publications (Rosental 2008). The diverse forms of demonstration are not always used in compartmentalized fashion, and, here again, the actors who plunge into demonstrative struggles can mobilize arsenals that are heterogeneous and dissimilar in nature.

If the demonstrative arsenals used in a multitude of social spaces appear highly diverse, then, the fact remains that they seem to depend on the sectors of activity and the skills possessed by the demonstrators. For example, to carry out its demonstrative actions, the NGO Greenpeace seems to privilege press conferences for putting forward scientific facts (e.g., measures of radioactivity, bringing to light a danger to the populations near a nuclear site), along with site occupations.

These phenomena, in an extension of those analyzed throughout the chapter, thus make it possible to grasp a certain number of limits relative to an approach to public demonstrations that would characterize them a priori as isolated acts. Analyzing the chains of action, and in particular the campaigns and demonstrative arsenals in which the performances are apt to be inscribed, appears essential if we are to apprehend their nature and stakes. I have in mind especially the way some demonstrators manage their demonstrative investments collectively over time and thus succeed in benefiting from the cumulative interest of their diverse public demonstrations. The campaigns they lead permit them not only to reinforce the credibility of their affirmations or their projects, but also to adapt them better to their audiences and to involve those audiences. They also offer demonstrators the possibility of building demonstrations of strength whose effects can be measured only over extended stretches of time and space.

We have also seen how demonstrators can be led to mobilize different forms of demonstration, selecting and combining them according to the context. The corresponding choices stem from a sense of which forms of demonstration are appropriate in a given situation, a sense that can be highly developed in certain individuals or within certain organizations. The constitution of demonstrative arsenals may for its part entail reuses, transformations, adjustments, and significant exchanges of demonstrations.

The scope of the demonstrative resources mobilized nevertheless does not represent a guarantee of effectiveness. We have seen that demonstrators may be involved in demonstrative battles and confronted with more or less imposing counterdemonstrative campaigns and arsenals. Such conflictual elements have to be taken into account if one seeks to pinpoint the effects that demonstrations are capable of producing on the public in the short or long run. However, many other elements also need to be analyzed to that end.

In particular, it is important to pinpoint the potential heterogeneous constraints and conditions that surpass the strict framework of performances, and to grasp in what respects the nature of the demonstrative devices in play and their uses are not the sole determinants of public reactions. Demonstrations can in fact produce nonhomogeneous, variable, or mitigated effects, and can even end up in more or less significant "failures," depending on the composition of the audiences addressed. These diverse reactions are managed in different ways by the actors before, during, or after the performances. These are the phenomena we shall examine next.

5

PERFORMANCES WITH LIMITED POWER

A great many elements have to be analyzed if we are to grasp what goes into public demonstrations and what their outcomes may be. As we have already observed by studying the ways these demonstrations can be prepared, combined, and exploited, some data cannot be captured simply by studying the way the performances themselves unfold.

To be sure, the talent displayed by a demonstrator and the content of the interaction that takes place during a demonstration play a crucial role in the way the demonstration is received.

During Microsoft's 2015 TechDays, for example, I was able to see how public presentations and demos produced audience reactions in response to the demonstrators' skills. During a plenary session, one demonstrator, speaking before a very large audience, had nothing like the ease of a Steve Jobs. He seemed to be reciting his text with the help of a prompter. While the delivery was fluid, for many listeners the tone was unconvincing. Audiences showed more enthusiasm when demonstrators were more adept. Similarly, at the 2014 Paris Fair I observed how skilled demonstrators succeeded in selling a large number of their products in their stands, while others struggled to sell even a few items of the same type.

Nevertheless, as a general rule, the outcome of demonstrations is not decided entirely on the spot. At the Paris Fair, for instance, the demonstrators' ability to win over potential customers depended not only on individual talent but also in part on the identity and predispositions of the audience. To grasp elements such as these, it did not suffice to study the conditions of interaction. In addition, the particular products being promoted during demonstrations at the fair helped predetermine the eventual success of the performances. Even with very gifted demonstrators, not every product sold well. Similar observations can be made about street protests. However skilled the militants of an association such as Greenpeace may be, various other elements help determine the effects of one of their site occupations, independently of the actual unfolding of events.

I should like to proceed, then, by bringing to light the importance of various "off-site" factors that determine the pertinence and the content of demonstrations, as well as their eventual outcomes. Again relying on several case studies, I shall examine in particular the extent to which the makeup of an audience and its reactions to a demonstration correspond to those targeted or anticipated by the demonstrator, and the extent to which these elements may be heterogeneous. I shall go on to analyze the way demonstrators anticipate or handle negative reactions, which are sometimes related to unexpected situations or malfunctions that occur either before, during, or after a demonstration is performed.

OFF-SITE FACTORS

Various off-site elements may influence the effects as well as the possible or actual content of a demonstration. To begin to develop this point, let us go back to the demonstrations carried out by representatives of a Hungarian bank for the purpose of selling home ownership savings plans. The scenarios of these demonstrations were defined essentially outside the framework of interactions with clients (Vargha 2011). Their content was created and modified over time by the bank's sales department, which took into account what had happened in earlier exchanges between its agents and their clients (219). Thus various types of demonstrations were adopted sequentially.

For a time, the bank's counselors used pie charts to illustrate the sources of funding that would come into play to finance the purchase of a home once the loan had been reimbursed, if the clients subscribed to a housing-specific savings plan. These graphic representations showed what proportion of the final sum needed to acquire property would be contributed by the client's savings, along with a state subsidy. Later, this stage of the demonstration was abandoned. In its place, the counselors were taught how to produce a standard demonstration that consisted in gathering information about the client's financial capacity, then making a calculation adapted to the circumstances, in order to show that the product offered by the bank corresponded to the client's specific needs. Demonstrators had to learn how to draw a graph by hand and comment on it as they went along, to make the pertinence of the proposed solution obvious.

As this case helps to show, the content of demonstrations is not always defined by the demonstrators themselves; it may be determined instead by other actors belonging to the organizations that employ them. Other constraints on the content may originate elsewhere: some may arise from the need to maintain public order, as we saw when we looked at the official limits placed on street protests. Preventing violence and controlling obstacles to traffic are considerations that affect the planning and shape of many street demonstrations.

The content of public demonstrations may also be subjected to legal constraints. In the context of the Delta project, for example, the demonstrators could not ignore the legal risks that would be incurred if they made exaggerated claims about their product. As they developed demonstrative repertoires, they had to keep in mind any potential dangers to persons or goods. Some participants in the project, for example, were concerned about the possibility that a catastrophic event in a space mission could be attributed to them, as designers of a computer program that could be used in the preparatory stages of the mission.[1]

Intellectual property rights were also taken into account in preparing demonstrations. The Delta project used software components under licensing agreements. When project participants were planning to carry out demos at the workplaces of their prospects, they had to make sure that the latter had the appropriate licenses on their computers; otherwise, adaptations would be needed (for example, remote connections) in order to carry out the demonstrations without breaking the law.

The Delta project demonstrations were subject to other types of constraints as well. Some were circumstantial in nature. The development of the earliest Delta demos was influenced by the crisis of credibility encountered by the American space program in the early 1990s. The public image of that program had been tarnished by a series of serious incidents, above all the in-flight destruction of the Challenger space shuttle in January 1986, a disaster that killed all seven crew members on board.[2] The catastrophe left its mark on the American psyche.

To help restore confidence in the American space program on the part of the public and political officials alike, it was important to show that the program was going to improve the reliability and safety of its equipment while also reducing its costs (McCurdy 2001). The demonstrative repertoires of the Delta project participants were thus clearly traced out: they had to emphasize above all that their software offered better performances and better safety conditions but also a reduction in the costs of preparing space missions.

Delta's demonstrative scenarios stressed the profitability and reusability of the existing software. They emphasized the automatization of calculation software production, which limited human interventions and "consequently" the risk of errors, while also reducing the costs associated with recourse to a highly skilled workforce.[3] These repertoires were all the more relevant in that they could easily be used in demonstrations targeting representatives of various industrial sectors. The possibility of reducing the costs of software creation and maintenance, coupled with the possibility of building secure programs, were elements that could be appreciated by many organizations (MacKenzie 1996).

While this case brings to light the way legal constraints and contingent circumstances can influence the content of demonstrations, we must also note the possible impact of moral considerations. Erving Goffmann (1974) mentioned situations in which demos should not be overly realistic, especially in matters of "decency," citing advertisements for bodily hygiene products in this connection. Alongside restrictions, we can also note, by contrast, practices and discourses that are appreciated by their targeted or actual audiences and that demonstrators might find especially useful or profitable in their presentations.

As an example, let us look at a demonstration of the hyper-instrument project that researcher Tod Machover developed at the MIT Media Lab. This apparatus was designed to produce computer-assisted music on the basis of virtually any type of object. To promote it in the United States,

Machover had created a demo showing how a tetraplegic individual could make music thanks to the hyper-instrument. As demonstrator, Machover made multiple gestures of empathy toward the musician, celebrating with feeling, in his commentary, the progress that had enabled the individual in question to triumph over adversity, and invoking additional values such as the solidarity of American society with handicapped persons.[4]

National political and cultural contexts must also be taken into account if we are to grasp the conditions that may affect the reception of public demonstrations. Several studies note that public demonstrations of technology tend to play an important role in the United States in efforts to justify scientific and technological choices to the general public and in campaigns to stimulate public confidence in science and progress (Merton 1938; Ezrahi 1990; Jasanoff 2005).[5] Such demos seem to be in harmony with a political culture in which individuals deem that they have the right and the ability to evaluate all sorts of public assertions and achievements, but also to debate them and even to produce competing claims. In a context where statements and accomplishments are constantly scrutinized and tested by citizens and critical media and where the expression of disagreement is an established way of producing public knowledge (Jasanoff 2005, 163, 270), demos offer handholds to private citizens and journalists alike in their efforts to assess the risks, costs, and benefits of the scientific and technological policies of a national government or of multinational companies, and they sometimes provide fodder for counterdemonstrations. In addition, public demonstrations of new technologies represent important tools for promoting research in a country in which science tends, if we can credit Tocqueville's observations, to be appreciated to the extent that it provides individuals with the means for increasing their material pleasures, augmenting their personal possessions, and reducing both time investments and production costs.[6]

Thus demonstrations of the utility of the scientific and technological research associated with the development of new weapons and new medicines during World War II seem to have played a more important role in justifying the creation of the National Science Foundation in the United States than the numerous arguments advanced to that end (Jasanoff 2005, 263). Similarly, the food industry in the United States has relied on several forms of public demonstration, including demos, to justify the development of genetically modified organisms (GMOs). In brochures and on websites, it has used images of immaculately laid-out fields to suggest how biotechnologies are triumphing over the disorder of nature and fighting world hunger. Public demonstrations have also put forward the idea that GMOs have been in use for many years without any evidence that public health has been affected (Jasanoff 2005). This demonstrative arsenal has provided clear targets for counterdemonstrations by opponents.

In comparative terms, science seems to benefit from greater public esteem in the United Kingdom, where it has not needed as great a stream of demos or indeed of any public demonstrations to win financial support and legitimize the choices made by the national government in standing behind the assertions of the scientific community (Jasanoff 2005). This seems to be the case in Germany as well. The confidence of German citizens in the scientific and technological choices made at the national level seems to be based above all on respect for procedures, especially the use of a deliberative process and the fact that there was consensus among committees of experts (Jasanoff 2005, 264).

These variations are elements that exert their influence a priori on the content and form that public demonstrations are to take, as well as on the more or less requisite character of such demonstrations, so that scientific and technological assertions, accomplishments, and choices will appear credible and legitimate in the eyes of the public, journalists, and business representatives (Jasanoff 2005, 249). This observation concerning science and technology can be extended to other areas, such as street protests. There too, international comparisons reveal that the more or less legitimate character of these actions as perceived by various types of actors, the nature of the actions, and the more or less frequent recourse to them in combination with or as substitutions for other forms of protest, can undergo significant variations and evolutions (Fillieule 1999).

While these tendencies as observed on a broad scale warrant our full attention, it nevertheless seems useful to take additional elements into account if we are to grasp the diversity of factors affecting the ways public demonstrations may be received at an individual level. A given public demonstration may in fact give rise to variable reactions from within the same audience, ranging from different forms of appreciation through mixed reactions to judgments of failure. The effects expected from a demonstration are not always achieved, nor are they always homogeneous. Neither a performance nor the devices it may bring into play can always determine reactions on their own, any more than they can construct an audience on their own. Let us look more closely at this point.

VARIABLE REACTIONS

If we are to believe Gustave Le Bon, demonstrations are not only ineffective for supporting assertions, they are even counterproductive. According to Le Bon, "affirmation pure and simple, kept free of all reasoning and all proof, is one of the surest means of making an idea enter the mind of crowds. The conciser an affirmation is, the more destitute of every appearance of proof and demonstration, the more weight it carries. The religious books and the legal codes of all ages have always resorted to simple affirmation. Statesmen called upon to defend a political cause, and commercial men pushing the sale of their products by means of advertising are acquainted with the value of affirmation" (Le Bon 2002, 77). This extremely general and radical assertion would need to be significantly nuanced, historicized, and localized even before we could envisage a potential demonstration based solely on empirical data. Still, it has the merit of providing an occasion to raise questions about the power of demonstrations and their possible involvement in the manipulation of crowds.

If scholars studying public demonstrations do not analyze the actual modes of their reception, either because they choose to focus exclusively on the unfolding of a performance without paying attention to audience reactions or to interactions between audience members and the demonstrator(s), or because they choose to focus solely on the concrete devices brought into play in this context, there is a non-negligible risk that they will attribute effects to the performance or to the devices that are more univocal and intense than the effects produced in practice. The actual effects are not necessarily aligned with the demonstrators' expectations or with the scripts that technologies put into play (Akrich 1992). Demonstrations, like technologies, can be subject to divergent interpretations and to deviations with respect to their inventors' intentions.[7]

The risk I have just evoked is encountered by historians of science and technology when they possess data on the unfolding of public experiments and on the machines used on these occasions, thanks to the existence of reports and iconographic sources, for example, but when they lack sufficiently precise data on public reactions. Contemporary sociologists of science and technology also run this risk if they focus essentially on the nature of the material equipment used in demonstrations—on what this equipment is presumed to permit and to constrain—without taking an interest in the actual interactions between demonstrators and demonstratees. In other words, we cannot settle for a study of the grammar of demonstrations, or for a materialist, anatomical approach to their equipment, if our goal is to pin down the effects they produce, and above all their power of persuasion.

The performances and associated equipment are all the less likely to determine audience reactions on their own in that audiences are not solely constituted by the enactment of demonstrations. In our earlier examination of demonstrations at a fair in Paris, we could already see that audiences might be formed of preconstituted groups, structured for example by family ties or friendships. Their members were thus not isolated and interchangeable individuals. They had socially "thick" features and bonds that marked their interactions during demonstrations. The nature of these audiences influenced the way the performances were received.

Moreover, the demonstrators at the 2014 Paris Fair were far from capable of controlling the composition of their audiences, even if they actively tried to gather spectators around their stands and sought to manage interactions with them. In addition to the families or groups of friends that turned up, the demonstrators depended heavily on the visitors that the fair as a whole had managed to attract; they were also dependent on where they were placed, since the client traffic varied in nature and intensity across the alleys linking and separating the stands. Similarly, the way spectators observed and interacted depended on numerous elements, such as predispositions they had acquired in the course of their individual biographical trajectories.

The experienced demonstrators at the 2014 Paris Fair were the first to judge that spectators' reactions to the demos they watched depended on their identities. This is why these demonstrators observed their audiences attentively and interacted with them in different ways in order to optimize sales. As previously mentioned, one demonstrator explained to me, for instance, that he looked for women of a certain age from the French Caribbean and counted on them to generate sales. According to him, their quickness to manifest a desire to buy his product had a viral effect on the other visitors clustered around the stand.

Other studies show that demonstrators at fairs in France dread the presence in their audience of a type they call *les barbus*, "the bearded guys" (Le Velly 2007). The label characterizes a category of individuals who manifest their skepticism during the demonstrations, asking multiple questions and insisting on detailed answers. Demonstrators generally spot them by the way they dress: these men often wear leather moccasins and a small shoulder bag. When a *barbu* turns up, the demonstrator sometimes decides just to end the performance.

"The bearded guys" are much more troublesome to demonstrators than the ones known as *renauds*, referring to visitors who come back to complain about the quality of the products they have bought (*renauder* is a French verb meaning to complain with ill humor). Experienced demonstrators have techniques for handling that sort of discontent: for example, they may explain to the complainer that he or she does not actually know how to use the product correctly.

Other cases help make it clear that the audiences for demonstrative performances are not always chosen by the demonstrators. The history of science has accustomed us to the opposite situation: demonstrators who control access to the demonstration site and determine who will witness it. Steve Jobs tried to create the same situation when he organized the launching of the NeXT computer in San Francisco's Symphony Hall (Elmer-Dewitt 1988). He and his team sent out a large number of targeted invitations with the goal of constituting an appropriate public for the event. One recipient was Bill Gates, the CEO of Microsoft. However, Gates declined the invitation in a public way, taking care to indicate that as far as he could see there was nothing revolutionary about the device. His presence would have brought credit to an innovation that could be in competition with Microsoft's own products.

Clearly, Steve Jobs and his team did not have full control over the composition of the audience for his demonstrative performance. Dynamics of competition that exceeded the context of the demonstration were in part responsible for the failure to create the intended audience. The decision of one particular invitee not to participate helped constitute Jobs's audience by diminishing it.

This case provides another illustration of the limits of studies that, seeking to grasp the outcome of a demonstration, focus exclusively on its actual performance, or, even more narrowly, on the device on display, without paying attention to the composition and reactions of the audience. The same argument can be made regarding the identity and attitudes of the demonstrator(s), as we can see for example in Christophe Traïni's study of the emotions expressed by the participants in anti-bullfight street protests in France in the 1990s. Going beyond a purely interactionist approach to these phenomena and analyzing the biographical experiences of the demonstrators from a broad sociohistorical perspective appeared indispensable for grasping the extent to which the emotions in question were not simply simulated for strategic reasons, and for understanding the effects of these emotions (Traïni 2010).

As a general rule, the presumed or targeted audiences for demonstrations and the presumed demonstrators have to be clearly distinguished from the actual demonstratees and demonstrators. A number of creators of high-tech demos imagine that their performances are addressed to a universal public, whereas in fact they are adapted to specific audiences that they themselves are familiar with or that resemble the creators themselves, typically educated representatives of the middle and upper classes. "Minimal USA," a demo in video form, offers a good illustration of this phenomenon. Presented in 2013 at the Museum of the City of New York in an exhibition titled *Making Room: New Models for Housing New Yorkers*, it featured a model apartment making the most of limited space.[8] The model, supposedly designed for New Yorkers in general, was in fact far from universal in scope. It corresponded to quite particular individuals and lifestyles; it was of very little relevance to blue-collar workers or immigrants, for example (Bean and Rosner 2013).

The reactions generated by such demonstrations can be all the more diverse when the demonstrations are confronted with a more heterogeneous audience than the one imagined or sought. The real audience can be made up of members whose varying identities and competencies do not always correspond to what the demonstrators had in mind.⁹ Cases in which the demonstrators manage to impose a homogeneous way of perceiving their demonstrations thus must not be viewed a priori as the rule.

To be sure, the history of science has shown, for example, how demonstrators of Newtonian theory in Georgian England succeeded in controlling the way their demonstrations were to be viewed, thanks to the discipline that they imposed on their machines and on themselves (Schaffer 1994). The multiplication of such constraining demonstrations may also help explain the emergence of a "paradigm" such as the "Newtonian paradigm."[10]

However, while this insistence on discipline may have a certain validity in cases such as those Schaffer describes, it is important not to plunge into hasty generalizations and imagine that constraining phenomena of the sort can be found under any and all circumstances. I have already pointed out how audience members at other types of public meetings—presidential visits, for example—have often been represented as sharing the same attitudes and viewpoints, whereas these latter were in reality quite diverse (Mariot 2001). It is also the case that some demonstrative undertakings, instead of organizing strict control over the way a presentation is seen, adapt to the diversity of viewpoints and proceed in a supple and gradual fashion.

I saw this process in action during my study of researchers in artificial intelligence in the United States (Rosental 2007). Demonstrations carried out at relatively early stages of a project were used to test audience reactions, which were solicited and analyzed in a procedure not unlike the one Thomas Edison used to electrify America (Bazerman 2002). By adjusting the projects and repertoires in a dialectical manner, one step at a time, these researchers were seeking to develop common viewpoints with demonstratees in the framework of performances.

Various historical studies describe the way scientists have tried to impose viewpoints by way of spectacular demonstrations, for instance in seventeenth-century France and England (Licoppe 1996). The term "spectacular" is often used by analysts to characterize demonstrations that have a theatrical dimension and are in some way impressive, striking, in the eyes of their viewers (Cossart and Taïeb 2011, 138). Yet the variability in viewers' reactions concerns the very notion of spectacularity. To the members of a given audience, the same demonstration may appear spectacular or not, or spectacular to varying degrees.

The case of the double pendulum examined earlier provides a good illustration. For researchers in space engineering, a demo of an automated mechanism capable of controlling the back and forth movement of a pendulum generally tends to have a spectacular dimension. Let us recall that such specialists generally see this mechanism as a solution to the canonical problem of a rocket at takeoff, which can break in two if the oscillations are too strong. Nonspecialists who do not make this connection are usually less impressed (Rosental 2004). Similarly, we can consider the case of logical demonstrations, whether published or written on a blackboard, that may appear astonishingly clever and thus spectacular to certain readers or audience members, whereas they do not appear in the same light to others (Rosental 2008).

These examples underline the fact that, even though many works in the history of science and technology bring to light the production of scientific spectacles, the expression "spectacular public demonstration" is not a pleonasm (Dolza and Vérin 2003). They also illustrate the fact that a given performance on its own does not necessarily determine audience perceptions of its degree of "spectacularity" in a uniform way.

As I have suggested, having access to enough data to grasp the variability of modes of reception of demonstrations is not always a possibility for historians. For sociologists working with contemporary data, by contrast, that task can be accomplished in a number of different ways. As examples, we can consider a series of demos carried out during Microsoft's 2015 TechDays.

Some of these demonstrations were produced on the stage of a large room in the Palais des Congrès, a major convention center in Paris, during an introductory plenary session before hundreds of spectators. They were also broadcast live on the internet. Others took place in smaller rooms, but still before a large audience, whether on site or via the internet. The public was largely made up of computer scientists and "geeks" equipped for the most part with smartphones, exchanging messages—tweets in particular—as they watched the performances.

The demo presented below illustrates one way in which this type of performance could unfold.[11] The text is a transcript of the first part of a

PERFORMANCES WITH LIMITED POWER 145

demonstration of an internet browser that Microsoft was calling Spartan (it was eventually released under the name "Edge"). The demo was introduced and presented during the first plenary session of TechDays by three of Microsoft's technology evangelists, Pierre Lagarde, David Catuhe, and David Rousset.[12]

DEMONSTRATION OF THE SPARTAN PROJECT BROWSER

PIERRE LAGARDE: [The presenter adopts a mysterious tone; behind him on a big screen one sees the haunted house of a video game; one hears scary music, and crows' calls.] So to head toward our second mission, I propose to take a secret passageway. We're going to get right into the ambiance, because it's a little dark. And for guides in this secret passageway, actually, David and David are going to be our guides.

DAVID ROUSSET: [Two demonstrators, David Rousset (hereafter DR) and David Catuhe (hereafter DC) are standing behind a lectern, equipped with a wireless microphone, dressed casually (in T-shirts). They control what is displayed on the big screen from portable computers.] Thanks, Pierre. So I have the opportunity to pull in an American who's going to try to speak a little French.

DC: It's not going to be easy! [The "American" is in fact a Frenchman working for Microsoft in the United States; he speaks with the distinctive accent of southwestern France.]

DR: And he's coming today with an exclusive product, an exclusive demo. We have the most recent rendering engine from the Spartan project, so it's the Javascript engine and rendering engine with an interesting new feature that we're going to show you together. But David, could you guide me through this demonstration?

DC: [The screen shows a fantastic décor around the manor house, which is explored on screen as the demonstrator comments; in the background, one hears the sound of crows.] So, while I'm telling you about it, I propose to go a bit to the left. You'll see, we have some neat little tricks here. So we got lucky, and I think it's one of the rare times that's happened in France. We had to fight hard, but we got the green light, actually, from Microsoft, for the first time, to show you something that hasn't been seen anywhere else. In fact, what you have in front of you here is Spartan's rendering engine. . . . And in this rendering engine, you have, for the first time—that's right—for the very first time, you have support for Web Audio. This is one of the biggest innovations that we're going to make available for you very soon. . . . And so David here, I saw you, you were scrolling a bit ahead with the Gamepad, which is another novelty, since we're going to . . . support the Gamepad in the Spartan project engine. So the scene here was set up by Master Rousseau, I really need to make this clear.

DR: And I'm going to show you something, here I'm stuck in what's called a collider.

DC: Oh yeah, that's interesting!

DR: And I feel like starting this demo all over from scratch. [Laughter from both demonstrators.] So it's not a bug in the rendering engine, it's a bug from the guys who coded the 3D engine. [On screen, there's a little wheel turning that indicates downloading.]

DC: Right.

DR: And that means you and me. [Laughter from both demonstrators.]

DC: Yep. So here, anyway, you can see that it's live.

DR: [The manor house appears again on screen.] Yeah, it's live. So, look, the joystick, isn't it terrific for moving around in a 3D scene? . . . So what were you telling me?

DC: So I was telling you that actually with Gamepad you have Web Audio, so it's super, because on the Web you're going to have the same experience wherever Spartan is supported. We'll be able to have sound, with—this is interesting, I don't know if you can hear it well in the room, but the sound is geo-spatialized. So, here's an interesting thing, instead of going to the keyboard I propose to . . . go to the mouse. That'll let us click on a couple of nifty things.

DR: So don't be scared, because we did some testing with a few of our young interns, and they got scared at this point. Listen . . . that reminds me a little of *The Seventh Guest* [an interactive video game released on CD-ROM in 1993].

DC: Exactly.

DR: For those who knew it, the oldest ones. [The demonstrator clicks on a shield, and frightening cries are heard.] There's what scared people away.

DC: Right. So you can tour the house, because we have a couple of nice little Easter eggs in there. Uh, that's also going to let me—while you lead the tour, here, and while we can appreciate the graphic quality on a little PC. . . . There, you hear the piano. [The demonstrator clicks on one of the house's windows and a skeleton playing a piano appears.] So it's actually a skeleton player who's having at it on his piano. We're also going to the cemetery, to have a really good atmosphere.

DR: A pretty nice little demo, no?

DC: [The demonstrators continue to move around the manor house and a cemetery appears on screen.] Yeah. And then in this cemetery, we have Hamlet. [The demonstrator clicks on one of the graves and brings up a skeleton.] Good, just for the fun of it.

DR: He's dead.

DC: You can make him turn up here. We have Clippy, who's dead too [the demonstrator clicks on a second grave and brings up another specter]. OK? You all asked for it, but it's all good, he's been killed.

DR: At the time of *The Seventh Guest*, he would have been recognized.

DC: We were wondering, the cat . . .

DR: Schrödinger's cat—is he dead, finally, or not?

DC: [The demonstrator clicks on a third grave and brings up an animal skeleton.] Well, yes, apparently he's dead. And IE6 [Internet Explorer 6] . . .

DR: Ah!

DC: And IE6 is dead.

DR: Finally!

DC: Finally.

DR: And look, when we're going to kill IE6 [spontaneous applause in the room], we're going . . . we're going to click [the sound of a bell is heard].

DC: And look, the new hope [the demonstrator clicks on a fourth grave and brings up crows that swerve around in the sky, forming a big circle]. Well, I would have loved to show you the new logo for Spartan, but we're not allowed to. We were allowed to show you Web Audio, that's already something. So be patient, because it's coming soon.

We need not linger over the perhaps hermetic references to the universe of video games and software: the demo took place on the first day of the TechDays conference, when computer specialists and "geeks" were particularly targeted. The demonstrators used a vocabulary and a form of humor adapted to that audience, poking fun at themselves: jokes about the "inability" of the demonstrators to conquer the bugs or to speak English without an accent combined well with the "frightening" character of the scenes projected and the symbolic burial of Microsoft's flagship software, the browser known as Internet Explorer. The new product of a major international company was thus presented as the product of tinkering by a couple of buddies.

If the operational mode adopted departed from the recommendations of someone like Kawasaki, who had little use for demos conducted by a duo, it corresponded nevertheless to other norms we have encountered: the use of familiar expressions, a quest for forms of complicity with the audience, an effort to impress, in this case through sound and visual effects, and an atmosphere of fantasy. The demo also featured a selection of brand new functions (in particular the possibility of using a gamepad in an internet navigator and the availability of high-quality sound), even as it aroused expectations by announcing other innovations to come. Moreover, the performance lasted only a few minutes, with no time wasted. Finally, in keeping with another norm concerning the order of presentation, a second part of the demo, not transcribed here, shed light on the software's working principles, introducing a new programming language.

This performance, like other demos presented during the introductory session, received positive comments in the specialist press, as the following excerpt from one report suggests:

Under the guidance of its new DX [Developer eXperience] director, Microsoft France seeks to reassure us: the in-house development tools are up to the task of responding to the technological challenges of mobile devices and the Web of the future. On February 10 through 12, the company is demonstrating its know-how to developers on the occasion of the now traditional TechDays. . . . Finally, there was a demonstration of the 3D rendering engine of Spartan, Microsoft's future Web navigator. We really got a sense that Babylon.js, based on webgl and javascript, is the cherished baby of Microsoft France's internal development team and that they are proud of it. So proud that they dared to let slip in passing that IE6 is truly dead, a reminder met with warm applause in the crowded hall during the opening session of TechDays 2015.[13]

What were the actual reactions of the audience to this demo, and to the demonstrations offered during the introductory session and others? If facial expressions and some words exchanged in the rooms during the presentations could be studied, they would constitute a limited set of data about the modalities of reception. The reactions expressed on Twitter, by contrast, constituted a much richer corpus. The fact that tweets were archived automatically by the app made it possible to follow the thread a posteriori.

This is an interesting case from an ethnographic standpoint. To the extent that several voices were being heard at the same time, those of the demonstrators on stage and those of spectators on Twitter, it was important to conduct multimedia investigations, combining on-site observations with an analysis of the exchanges taking place on the social network. Limiting oneself to observation during sessions alone would be in a sense comparable to watching a film without the sound. Some of the participants' perceptible actions would not have been taken into account.

An analysis of the tweets makes it clear that the reactions to the demos were highly varied. The style of the tweets was determined in part by the identity of the intervenors and by the fact that these actors were expressing themselves publicly on a social network marked by a specific form of communication: at the time, for instance, messages posted on Twitter could have no more than 140 characters.

Certain tweets conveyed positive and often enthusiastic reactions to the series of demos presented during the plenary session. A few examples:[14] "Convincing demo of Azure websites and especially showing how simple it is to do testing in production!" "Damned fine team of tough guys on stage!" "Spartan project demo—exciting!" "Nice, the Spartan demo." "The Visual Studio demo is mouth-watering!" "Wow! Amazing demo of IoT using Visual Studio." "Not bad, the android emulation on Visual Studio, no?" "Very good demo by Sébastien Pertus." "Ultra powerful Dataviz DEMO through the creation of apps for #Office." "Good plenary Day 1." "Top-notch, this first plenary session. Bravo and thanks to the speakers :-)." "Day 1 plenary over! Full achievement: 3 stars! Great job guys." "Impressed by this first plenary of #mstechdays! Azure, Spartan, Visual2015. . . . A promising future!"

However, more equivocal or even openly critical reactions were formulated at the same time by other actors on the social network, even if the humor of the computer geeks continues to stand out. For example: "Azure demo not very interesting for those in the know." "Well, guys, you know, a white text on orange background, you really can't do this." "Presentation of the Spartan project . . . bugs already . . ." "Machine Learning at #mstechdays? Crystal ball mode, and crude test :)."

While the most positive comments emphasized the quality of the demonstrators and the demos, and sometimes their spectacular character, the reactions were clearly not unanimous. In particular, one writer who identified himself as a user of the Azure platform stressed a gap comparable to the one described previously for the demo of the double inverted pendulum, between the possible reactions to the demo of "naïve" viewers and those of the "in-group." Far from appearing spectacular to this writer, the demo struck him as of limited interest. The critical comment about the choice of colors used in a demo recalls the similar considerations expressed during the preparation of the NEO system demonstration.

This gradation of reactions appeared with other demos as well. In one case, a very popular session called "Coding for Fun," which gave the computer specialists present on stage the opportunity to create amusing programs, gave rise to very divided reactions. Some tweets were especially enthusiastic, using expressions such as "cool," "powerful!!!," "How to put it, heavy stuff." Also: "Even more music!" "Not bad, the musical intro of #c4f this year." "The sound is terrific!" "Top intro." "Encore!! Encore!!" "Megalo demo!" "Demo live really fun, tops as usual!" "Classy! Very very impressive as usual :D." "Hot damn, it's a killer demo!!!" "Code, demos, demos, fun, demos, code #c4f ;)." "Bravo to the #c4f team! I laughed so hard I cried." But other messages had a much less complimentary tone, as in the following tweets: "A little too well-rehearsed, this session. The beginning is really too cut-and-dried, too organized." "The intro is just like the rest . . . a big whatever." "The demos are fine. With the Internet it's better :)." "Well, this time, there isn't going to be good press. #tadeo No professionalism. Left us high and dry."

Some of the tweets just cited emphasize the spectacular character of the demos. Most notably, they express positive reactions to the sound effects and the humor displayed by the demonstrators. But others formulate critical judgments by stressing technical problems and failures, or, conversely, overly slick demonstrations.

If we analyze the full set of tweets formulated during TechDays, we note that, as a general rule, positive judgments bearing on the demonstrations and their authors (e.g., "The best demo of #mstechdays up to now is the HTTP2 by @deltakosh and @davrous." "Very nice, the demo of #cortana by @dupujs and @Caroc." "Fine demo of VSO and the integration of the hook services!" "Nice, the #minecraft4dev demo.") coexisted on the social network with negative judgments that were often more targeted. Some honed in specifically on the technical or visual aspects of the demos (e.g., "he didn't even personalize the ppt. ms . . . the 'cut-and-paste' bit at the bottom of the page is pathetic." "Cortana doesn't understand and insults a speaker during the demo! Still more progress to be made"). Others concerned the performances and comments of the demonstrators, as in the foregoing examples. Still others had to do with the commercial character of the demos (e.g., "Session on migration exchange to office 365—but no tech at all! It's a binary tree commercial presentation!!" "Crummy presentation . . . just sales prep."). The marks of disapproval corresponded to the register of anti-capitalist critiques formulated frequently by certain hackers (Auray 1997).

Microsoft TechDays is of course hardly the only context in which it is possible to produce evidence of the variability of reactions to demonstrations. The debates to which the demonstration of Charles Elkan's theorem gave rise offer another illustrative example. The study of messages addressed to a specialized electronic forum brought to light highly differentiated reactions to Elkan's demonstration, as well as to the demonstrations and counterdemonstrations produced by various interlocutors.

In reacting to the reactions, certain participants manifested a strengthening of their viewpoints on the theorem, as well on its demonstrations and counterdemonstrations, while others changed their views. However, upon reading these same messages a number of participants quickly settled into viewpoints based on antagonistic normative approaches to what logic, or fuzzy logic, or a logical demonstration, or the ways of debating any of these, *ought* to be. At stake in particular were sometimes highly ingrained notions of what should be prioritized in writing a demonstration like Elkan's, that is, how a proof should be organized and what elements were required to make it correct at each stage.

Formal proofs, of course, offer users guides as to how to read them. But owing to the heterogeneity of conceptions concerning the exercise of logic, and to the corresponding heterogeneity of skills—beyond a minimum of shared competencies—in the participants in the debates over Elkan's theorem, these participants generally did not read the arguments the way their authors had intended. This gap also stemmed from the fact that the demonstrations and counterdemonstrations were polysemic: even though they used symbolic languages, they were open to divergent interpretations, resulting in widely scattered and contradictory viewpoints on validity of the theorem and its demonstrations.[15]

This case provides yet another illustration of the variability of the reactions that public demonstrations can arouse, and it also shows the extent to which these reactions are linked to the identity and competencies of the demonstratees. It must be noted, however, that this variability does not always represent a pitfall for the demonstrators, especially if they themselves are ambivalent about the status of their performances. Indeed, demonstrators may sometimes actively seek this variability. The famous case of exhibits of "mermaid mummies" is an illustration of the phenomenon.

In the mid-nineteenth century, the showman Phineas T. Barnum rented what was often presented as the mummified corpse of a mermaid and exhibited it in the museum he had created in New York ("The American Museum").[16] But his goal was not to convince everyone that mermaids had actually existed, by bringing proof of their reality via the display of a specimen; in this he distinguished himself from some of his competitors. One such competitor's poster featuring a mermaid exhibit was unambiguous: "Whereas many have imagined that the history of mermaids mentioned by the authors of voyages, is fabulous, and only introduced as the tale of travellers, there is now in town an opportunity, for the nobility, gentry &c. to have an ocular demonstration of its reality" (Vidal 2006, 71).

Barnum chose, on the contrary, to invite all visitors to examine the object on display in his museum and to judge for themselves whether or not it was authentic; it was up to each viewer to decide whether or not the exhibit demonstrated the existence of mermaids. Not only did he refrain from taking a clear stand himself on the question, but he did not hesitate to stir up doubts on the subject in his public declarations. By bringing to light contradictions in their arguments, he sought above all to disqualify the viewpoints of scholars who denounced the object as a fake. In addition, he sharpened the spectators' curiosity and flattered their perspicacity by explaining that they themselves were capable of distinguishing what was real from what was fake by making direct observations and exercising their critical faculties.

Confronted with a very well-made object whose exact nature was hard to discern at first glance, visitors to Barnum's exhibit reacted in a variety of ways. The divergence in their understanding, like the maintenance of doubt in many visitors' minds, was exactly what the showman was after. His main goal was not to obtain a convergence of viewpoints on the validity of a proof (in this case, proof of the existence of mermaids) but to attract curious spectators who thought themselves capable of making their own judgments without relying on the opinions of specialists; such spectators would want to do their own authentication while being entertained by the sight of an unusual object. From this standpoint, the uncertainty that was maintained, the perplexity of some viewers, the dissent and even the polemics in the face of this display were important assets for Barnum. They were in fact sources of publicity, apt to interest and attract new clients.

The mermaid mummy case is interesting on more than one count. First, it highlights, as did the demos of the European ACTS program we examined earlier, how audience members can be invited to forge their own opinions on various phenomena on the basis of performances, and how the latter then represent an alternative to reliance on expert opinion. Next, it shows how the status of public demonstration—understood in the strong sense, as public proof—is not always attributed in a straightforward manner, either by the spectators or by the demonstrator, to a given performance. Overtly expressed doubts and challenges may emanate from the originator of the demonstration as well as from its observers. These attitudes are interesting to study if the objective is not to arrive at a definitive judgment as to the status of a demonstration, but rather to understand how that status can be attributed to a performance, or not, in a collective way.

The mermaid mummy story also shows that the effects that can be programmed into a demonstration, whether its purpose is clearly demonstrative or ambiguous on this point, are not always homogeneous in reality, especially if ambiguity is the goal sought, and if the demonstrator is particularly skillful in this respect. But it is important to note that a demonstration can even arouse audience reactions that are the opposite of those sought, especially if the demonstration does not go as planned. Let us consider examples of such situations and observe how demonstrators may handle unexpected incidents before, during, and after their performances.

DUDS AND BUGS

Demonstrators usually go to some lengths to avoid "duds," trying especially to eliminate bugs, or at least to handle them as well as possible when they occur. Still, failures with respect to the goals set for the demonstration cannot always be avoided. Moreover, the border between a public demonstration deemed a "success" and one written off as a "failure" is not entirely defined by the way the performance goes. It also depends in part on the variable modes of apprehension on the part of the spectators, and on the stories that are produced after the fact by the demonstrators and by the demonstratees. Some examples will enable us to examine these points more closely.

First of all, we must note that a demonstration, even if it unfolds as anticipated, may not simply give rise to mixed or ambivalent reactions; it may completely misfire in relation to the demonstrator's expectations. The case of a demo carried out in Paris in 2015 in the office of Minister of Ecology Ségolène Royal, intended to influence industrial policy, offers a good illustration of this phenomenon.

For a number of years, the Alteo factory that produces aluminum oxide in Gardanne (Bouches-du-Rhône, France) has been dumping toxic wastes such as mercury and arsenic into the Mediterranean, near the rocky cliffs of Cassis. The wastes are transferred by means of a canal system several dozen kilometers long that feeds into an underwater canyon a few kilometers from the coast. These wastes take the form of red mud. Following an agreement to protect the Mediterranean signed by France, Alteo was required to propose a treatment for the waste products before releasing them into the sea. To show that the company was planning to comply with the goal of zero waste, its representatives developed a demo intended for Minister of Ecology Ségolène Royal. Here is a newspaper account of the presentation (Thénard 2015, 3):

> The executives of the Alteo factory in Gardanne have shown real pedagogical talent in demonstrating the effectiveness of their anti-red-mud depollution plan. A few months ago, they were invited to the Ministry of Ecology, on the Boulevard Saint-Germain in Paris, to present their new technique to Ségolène Royal in person. They had brought with them a small case containing a little vial. The head demonstrator held out the flask for the minister to smell. Inside, water that was not red but transparent. "This is proof that what will be discharged into the sea from now on is completely harmless," he explained. "Well then, if this water is harmless, drink it," Ségolène replied. The onlookers were stunned. The demonstrator was unwilling to drink it. So were his colleagues. Proof that there is something fishy in the water?

This account illustrates the way a demo can completely miss the mark. The display of transparent water was supposed to be a decisive demonstration of the effectiveness of the company's treatment of its red mud wastes. This is a classic approach; as we have seen, the effectiveness of a degreasing product was demonstrated in a very similar way at the 2014 Paris Fair. The crowning touch of that demo consisted in showing how a glass full of viscous liquid became colorless in contact with the product.

The effectiveness of this type of demo and its spectacular character are not obvious properties for every audience, however, as the minister's reaction shows. Through her remark and the manifest apprehension of the demonstrator and his colleagues at the idea of drinking the liquid on display, Ségolène Royal managed to contribute a counterdemonstration, or at least to cast doubt on the initial claim that the waste products of the Alteo factory would be completely safe owing to its treatment protocols.

This case provides additional evidence that a demo does not systematically control the way it is perceived, nor does it automatically annihilate the critical sense of the viewers and turn them into mindless enthusiasts. On the contrary: various elements, some of which exceed the framework of the interaction, may play a decisive role, elements such as the identity of the audience members, or their preexisting, independent knowledge of the issue.

The unexpected reversal encountered by the Alteo representatives in their attempt to demonstrate their claim is hardly an isolated case: similar situations can be observed in many other circumstances, as I was able to see for myself during my investigations at the Paris Fair. At certain stands featuring rather unskilled demonstrators, the effects sometimes fell flat: spectators rapidly exited through the shifting circle of people around the demonstrator, and few sales if any resulted from the demos. Michel Audiard's portrait, in the 1974 film described earlier, of a hapless salesman-demonstrator whose sales techniques all fail,[17] thus is not simply a fiction.

To the fiascos created by unanticipated, hostile reactions from the audience, even when a public demonstration is performed according to the script, we can add those that stem from unexpected problems, and especially from technical malfunctions that occur during the performance. Bugs can have catastrophic effects.

To be sure, demonstrators sometimes actively seek them. Certain hackers take pride and pleasure in exhibiting their virtuosity by showing how quickly they can handle bugs and even total crashes, right in the middle of a demo. Such seemed to be the case with Apple's Andrew Hertzfeld, as we have seen (Lunenfeld 2000). Some programmers who attend demos give little credit to performances like this when they fail to encounter any bug, deeming them "too well-rehearsed" or "too slick." In this way they are denouncing the tedious or emblematic character of the marketing methods proper to a capitalist system that they are critiquing (Auray 1997). Among the tweets posted during Microsoft's 2015 TechDays that I analyzed earlier, there was a good illustration of this posture. Its author said, in effect: "A little too well-rehearsed, this session. The beginning is really too cut-and-dried, too organized." It is clear, then, why the Spartan project demonstrators might well insist on staging a bug during their presentation at the TechDays introductory session. Facing an audience of programmers, improvisations and bugs such as the ones that came up during several demos can actually be sought and welcomed.

In addition, some demonstrators are aware that, when a commercial demo comes off without a hitch, it can appear suspect in the eyes of the demonstratees, who may feel that they are being shown an arrangement specially designed for the occasion, like the presentations of "vaporware" products we discussed earlier. In such cases, too, the demonstrators may seek to avoid producing demos that go too smoothly.

In the realm of electronic games, moreover, there is an audience partial to bloopers; this is why players make demos revealing the existence of bugs in gaming software, when they come up during actual games. These demos, as they are recorded, strung together, and posted as videos on the internet, are highly popular among gamers. One such collection, "The Top Ten Worst Bugs in Video Games," had been viewed more than 340,000 times by the fall of 2017.[18] In this video sequence we see a character who is strolling along although he has lost his legs, a soldier who is moving about with a sword planted in his head, a horse that becomes its own twin, an individual who speaks after dying, a tank that moves through the air, and a man on a jet ski plunging ahead even though his machine remains immobile.

Certain videos featuring bugs that have come up during demos carried out in the context of product launchings are also quite popular. For example, an excerpt from a demo performed by Steve Jobs to launch the iPhone 4 includes a technical glitch; it had been watched more than 1.6 million times by fall 2017.[19] Given this attraction to bloopers, bugs may actually constitute a source of publicity for products. In this context, bugs may be sought after, or at least accepted, by both manufacturers and demonstrators, as long as the latter succeed in handling them appropriately.

If malfunctions can thus be valorized in certain cases, we must remember nevertheless that demonstrators generally do try to avoid them, and they resort to various measures to that end. In keeping with the advice of tech evangelist Kawasaki, certain demonstrators take care, for example, not to let the performance of their demo depend on a Wi-Fi connection, and they make sure to have two or more sets of equipment available. Others, as in the case of the Delta project, prefer to use frozen versions of the software under development—not the most recent or the most sophisticated version, but versions with whose operation and defects they are intimately familiar.

For an IT service provider presenting a demo to a prospective customer, or in the context of a call for bids, a major bug can in fact have disastrous effects if it cannot be circumvented. The potential client often is ready neither to organize a new meeting involving members of one or more teams, nor to run the risk that what she has observed will be reproduced in the everyday operations of her company (Smith 2009); a number of cases have been documented in which the candidacy of an IT service provider has been rejected following a "failed" demo.

Beyond the commercial failure that it can cause, a major bug can also have negative consequences for the demonstrators, from both professional and moral standpoints. The demonstrators' honor, dignity, and credibility may be affected, along with their self-esteem and pride in relation to their colleagues and clients, if the bug turns out to be memorable. Even as they aim for commercial success, demonstrators may also seek the satisfaction of a job well done (Sherry 1998). Thus a bug can simultaneously undermine their own self-confidence, tarnish their reputation, and have a negative effect on their professional future.

In his advice to start-up developers, Kawasaki does not hold back on his warnings; he has especially harsh words for demonstrators whose demos fail owing to technical deficiencies attributable to a lack of preparation.[20] In this respect, he is no more tolerant than was the Royal Society in the seventeenth century (Shapin 1988). Kawasaki is unsparing in his moral condemnation of demonstrators who fail to bring along several sets of the requisite equipment, who do not carry out all the appropriate verifications, or who take careless risks, for example by responding to questions from the audience during their performance. He speaks of "zero slack for equipment failures," and he calls the behavior of those who fail to take these precautions "clueless." He makes it clear that if a demonstrator is "mediocre," "the whole world" will know.

This sort of statement highlights the moral and professional sanctions that can accompany the appearance of bugs in demos, especially if the demos are not well "managed" by the demonstrators during their performances. For numerous strategies can be brought to bear.

We have already seen some of Gallo's advice for confronting such circumstances. According to that communications consultant, demonstrators must keep their cool, draw the spectators' attention away from the bug by recounting an anecdote, and mention what should have been shown before passing quickly to the next point. During a talk at

Microsoft's TechDays 2015 devoted to mobile devices, the demonstrator did just that after encountering a bug in his demo. The demonstration was addressed chiefly to an audience of computer specialists, so it drew on technical considerations beyond the competence of nonspecialists (we need not go into them here, for they are not essential to the analysis). Introduced by Nicolas Petit, marketing director of Microsoft France, the presentation consisted in showing, in less than six minutes, how a piece of software designed for one of Microsoft's operating systems could be converted for use in another Microsoft operating system. Dressed in a T-shirt, demonstrator Rudy Huyn stood behind a lectern and carried out manipulations on a portable computer and a smartphone that were projected on a big screen. An excerpt follows from that demo, purged of some of its more technical passages.[21]

DEMO: "UNIVERSAL APPS"

NICOLAS PETIT: Whoever thinks about universal apps thinks about demos, and whoever thinks about demos thinks about Rudy, who's going to help us discover some things about universal apps. [A text on screen reads "Demo: Universal Applications. Rudy Huyn."] Rudy, it's all yours.

RUDY HUYN: So, thanks anyway for that introduction. OK, . . . I'm going to tell you a bit about universal apps. . . . I started with a little code that I worked out for this occasion, it's called *Tweet Days* [the demonstrator carries out some maneuvers on a mobile phone that are projected on the big screen]. So I can show you the app quickly . . . every two seconds it displays very briefly the tweets related to MS TechDays. So, roughly speaking, it's a list with a pivot, buttons that go to launchers, options, other pages, and so on. . . . So . . . what I'm going to do [the demonstrator makes some moves on his computer and the content is projected on the big screen] . . . is take this app that I made in WinRT, and try to convert it while you watch into a Window 8 app, and show you how simple it is. . . . Now here we're going to try to see what you get in Windows. But I'll warn you, it's not . . . it's not a *fail* on my part, it's going to . . . it's going to . . . it's going to crash, and that's normal. So why does it crash? The first thing we're going to check is that . . . we have a little bit of code. There we are! There are even . . . there are lots of codes here, for what it's worth. [Laughter.] I must have left out some little piece. Let's see.

Uh, . . . OK, there I have a file that's shared by the two solutions. . . . So we're going to try to compile this and see why it doesn't work . . . Ah, OK, I made a beginner's error, it's my fault. When I created the Windows project I forgot to get rid of the little files it puts in there and that I don't need. Here we are, it ought to work better now. So, my Windows 8 project, now I can run it. And have a little error! [Laughter.] Well, this one, I think it was done on purpose. . . . Because if I keep my shared code, on my page, I realize that I'm using a pivot. . . . So that's not enough, because, actually, the hub has its own logic, somewhat virtualized. . . . Uh . . . [Laughter.] Ah, OK, I promise that it's working. Now this . . . it can't be done. It's not serious, we're going to trust the designer. Fine, it doesn't want to, it's really a fail on my end. Uh, I just wanted to show you one little thing, it's that, potentially, my list view . . . I'm going to be able to replace it by a great view and have an app that will work without a hitch, and that will be quite nice graphically. There, so that, for the first demo, OK, sorry about the technical problem.

As I suggested earlier, the fact that this demo does not appear "too well-rehearsed" had several advantages for seducing the audience of computer geeks to which it was addressed. While it put an "intentional bug" on display, aiming to show the effectiveness and simplicity of the arrangements for converting software, it also ran into an unintentional one. The demonstrator adopted an attitude in keeping with Gallo's advice. He dealt with the bug in a humorous and relaxed manner. Unable to get around it quickly, he sought to de-dramatize the situation by choosing to laugh about it. Confronted with an impasse, he did not linger over an admission of failure and did not stubbornly try to find a solution. He explained what the public ought to have seen before making a quick apology and moving on.

Looking far beyond this specific case, ethnomethodology has shown that when a malfunction interrupts the ordinary course of any interaction at all, the actors generally find it important to reestablish the intelligibility of the situation (Garfinkel 1984; Capelle 2012, 161). However, this effort does not necessarily entail lingering over the reasons for the malfunction. Passing over such an interruption quickly may be a virtue in the demo presentation.

Another strategy demonstrators use to confront this type of situation consists in taking verbal precautions at the outset. If the demo features a product that is not yet on the market, announcing that a "prototype" or a "version 0.1" is going to be presented and that the device or software therefore is not necessarily bug-free can make a malfunction more acceptable.[22]

In addition, announcing that a bug may appear during the performance can be a way of emphasizing the revolutionary character of a technology, even as it helps to establish a certain complicity with the audience. We have already seen how Steve Jobs used this strategy when he was launching the NeXT computer. By evoking the possibility of a bug in a humorous tone and calling for the viewer's compassion, he could hope to accomplish several things at once: highlight the merits of the technology, institute a certain complicity with the audience from the start, and solicit indulgence when any problems arose, having demonstrated a certain well-mannered humility.

During my investigations I have come across a number of demonstrations whose authors showed a sense of humor on the subject of possible bugs and on their own supposed limited abilities as demonstrators. This tactic often hit the mark. We have already seen how demonstrators joked on this subject during Microsoft's 2015 TechDays. This attitude showed up even in tweets; one demonstrator said in effect: "My demos always fail #mstechdays@davrous." As we have seen from other tweets, the rigor behind the production of fluid, efficient, and flawless demos is often such that by contrast this sort of assertion can come off as humorous.

We should also note that a bug can also be "managed" at the end of a performance. One possible tactic consists in trying to minimize its importance by explaining, for example, that the "version" of the device presented has been or will soon be corrected. Another possibility is to valorize the bug, by explaining, for example, that it is actually useful in the context of developing the device, learning how it works, or preparing demos for it. The opportunity to progress in the art of demos can also be stressed. This was the bottom line, for example, for the artistic directors of the NEO project, after deploring the fact that a colleague had made a bad last-minute decision to change the color scheme used in presenting the system without seeking the directors' approval.

There are in fact many ways to make a public demonstration come across as a "success" rather than a "failure," whether or not there have been malfunctions. Consultants in the realm of new technologies who are accustomed to producing success stories have lots of examples to offer. During a major international conference on information technologies, for instance, one consultant undertook a somewhat contorted exercise. He had sought in the past to demonstrate the "inexorable" character of certain predictions about the future of technology, predictions that later came to be considered totally erroneous. During the conference, he tried to demonstrate before an audience of more than two hundred people that in certain respects these early predictions had turned out to be at least partially accurate (Pollock and Williams 2010, 536, 537).

This example emphasizes that the possibility of declaring that a demonstration has met with failure or success is not determined simply by the content of the performance. Such a declaration can result from after-the-fact labeling carried out by actors who have significant margins of maneuver for supporting their assertions. The case I have just mentioned also illustrates the temporary character that such assertions may have; another clear illustration can be found in the variability and the evolution of judgments passed on public experiments at the Royal Society of London in the seventeenth century relating to the phenomenon of "abnormal suspension of water" (Shapin 1988).

This phenomenon, brought to light by Dutch physicist Christiaan Huygens, had given rise to major debates in the Royal Society in the early 1660s, after having been the object of public experimentation. In a first phase, certain eminent members of the society such as Robert Boyle had concluded that the phenomenon did not exist, and that what was seen during the experiment was simply the result of a leak in the apparatus used. After Boyle himself conducted the experiment and Huygens came to London to demonstrate the phenomenon before witnesses, the "failure" of the initial experiment was finally perceived as a "success." The members of the Royal Society acknowledged the existence of the phenomenon and judged, conversely, that the public experiments that had not brought it to light were "failures."

Judgments of the failure or success of a public demonstration thus benefit from being analyzed on more than one level, whether these be temporal, spatial, or interpersonal (Collins 1988). A multilevel analysis makes it possible in particular to connect the production and the relative durability of certain judgments with the identity and the interests of the actors who formulate those judgments, as well as with the sociohistorical spaces in which the actors evolve and with the dynamics of competition in which they are inscribed. The adoption of such a principle of variation also makes it possible to grasp to what extent this production of judgments may be determined by elements other than the content of the demonstration in question.

To take another example, street protests that took place in Newbury in the United Kingdom against the construction of a road did not put an end to the project (Rosental 2011). For the neighborhood, the situation represented a manifest failure. However, for certain organizations such as Friends of the Earth and the Green Party, which had supported the movement, the conditions under which the battle had been lost had helped reinforce their own efforts in combat being carried out on a much larger temporal and spatial scale for the defense of the environment against big construction companies. The latter had been weakened by the conflict. For the organizations involved, the street protests were not straightforward failures after all.

The struggles in France that pitted the urban community of Grenoble and the NGO Vivagora against the Pièces et Main d'Oeuvre association (PMO), bringing together opponents to projects related to nanotechnologies in the region, provide another illustration of the variability of judgments that may be rendered on performances upon the conclusion of demonstrative confrontations. According to one study dealing with these dynamics, the various parties had adopted different criteria for failure and success as they came to conclusions about the actions that had been taken (Laurent 2011, 2017).[23] For the administrative authorities and the scientists involved, the success of the citizens' debates over nanotechnologies that had been organized by Vivagora was measured primarily according to the degree of confidence in nanotechnologies that the public had acquired. For Vivagora, the judgment of success was based, among other factors, on the level of citizen participation in Grenoble. The positive evaluation made by PMO members, by contrast, was based primarily on the number of passive spectators who had been transformed into citizens critical of nanotechnologies. This example thus suggests that different declarations of failure or success can be based on divergent interpretations of these terms and can result from elements independent of the actual performances.

The foregoing example is far from unique. In the context of a street protest, the organizers, political representatives, and journalists frequently come up with conflicting assessments, evoking success or failure on the basis of quite different criteria. In France, for instance, labor unions generally congratulate themselves on the success of a demonstration they have organized, measuring success on the basis of their participant count, while the authorities report a much smaller number of participants and thus declare the event a failure (Champagne 1984). According to their own political inclinations, among other factors, journalists may deem a street protest a success, or they may stress the unhappiness of individuals who have suffered from the disorder caused by the event and emphasize the hostility of part of the population to the cause being defended. My own study of the 2014 theater professionals' street protests in Avignon and their media coverage provides one example among many.

Many declarations of failure pronounced with regard to public demonstrations seem to be shored up by new demonstrations produced by opponents of the previous ones. These opponents often have to show where the failure lay in order to use the term. Many declarations of success also seem to be the object of demonstrations. Thus it is common for the production of "successes" and "failures" to be inscribed within the context of demonstrative struggles.

We must note, moreover, that a declaration of success with regard to a public demonstration generally has to be distinguished from the achievement of the objectives that may have motivated that demonstration. For example, Steve Jobs's demo of the NeXT computer seems to have been received unanimously as a success. Journalists, analysts, investors, and even the competitors cited in the media all expressed enthusiasm after the performance. However, sales of the new computer fell far below expectations (only 50,000 were sold in seven years). The company that manufactured the product was finally bought by Apple, less out of interest in the computer than for the possible applications of some of its elements in the development of the internet (Lampel 2001).

Cases of public demonstrations perceived initially as successes that were followed by similar unsatisfactory results are numerous. As one example, we can look at a call for bids organized by a municipality in the United Kingdom seeking to set up a computerized customer relations management system (Pollock and Williams 2010, 534). Most members of the selection committee had been favorably impressed by the demonstration presented by one particular IT service provider. However, that candidate was not retained by the committee in the end, on the basis of arguments presented by a consultant whose advice had been sought.

These last two examples provide further evidence that the power of public demonstrations has to be evaluated on multiple spatiotemporal scales if we want to avoid crediting these performances with effects that they do not have. We are now better positioned to see why it is important not to limit the analysis to the mere observation of the performance as it unfolds.

We began by noting that the reception of demonstrations and the definition of their content are subject as a general rule to all sorts of constraints, which may be organizational, legal, conjunctural, or related to various political or cultural configurations, or a combination. Moreover, we have observed that neither the performances nor the various material devices they may mobilize determine on their own the composition of the audience, the reactions of audience members, or the nature of the interactions. In the cases we have examined, all these outcomes appear quite variable, and in some case quite different from the results sought. We have seen that, to try to limit the volume or degree of negative reactions, especially those that might result from unanticipated glitches or malfunctions, demonstrators may fall back on a variety of strategies. Some of these entail precautions taken prior to a performance; others have to do with techniques for managing unexpected incidents (humor, restoration of the intelligibility of the situation, quick transition to the next sequence, and so on). Still others entail a posteriori valorization of problematic episodes, in an effort to transform what appeared to be a "failure," at least to some observers, into a "success." It turns out that demonstrators and their critics have a more or less ample set of tools and margins for maneuver that they can use to shore up such contradictory judgments.

Public demonstrations thus seem to be endowed with a power that is limited in several respects. But even though they are not all-powerful, their effectiveness appears to be far from negligible in the various cases we have encountered. As for the demonstrators themselves, they appear to be neither totally powerless nor endowed with absolute power by virtue of their tools. However, in certain contexts, specific skills that they have managed to acquire, most notably regarding ways of dealing with bugs, seem particularly influential in judgments rendered on their performances.

We may well ask, then, how demonstrators acquire the various tricks of the trade. Is the mastery of such skills typically inscribed within the context of apprenticeships, or are they simply competencies developed within a given sector or profession? We shall consider these questions in chapter 6. Here again, it is a matter of getting beyond a narrow framing that would be based solely on a series of observations of distinct performances as they unfold; we shall seek to gather a larger number of elements that will help us grasp the conditions of possibility and stakes of these events. In bringing elements of response to these questions, my goal is to better circumscribe the nature of the practices in play, and above all the resources mobilized in the conception, realization, and accompaniment of public demonstrations, along with the constraints that bear on such exercises and the conditions that make them possible.

6

APPRENTICESHIP AND PROFESSIONALISM

In previous chapters, we have seen various specialists in marketing technology offering advice on how to prepare and deliver demos in articles or books, on the internet, or in oral presentations. Are these the only resources available for imparting such instruction? Or is there an organized system of apprenticeship in professional contexts, and if so, what processes and skills are involved? In order to bring some initial and at least partial answers to these questions, I shall begin by looking at the way skills are acquired in the realm of artificial intelligence research for the purpose of organizing demonstrations, within trade unions for organizing street protests, and in the pharmaceutical industry for organizing promotional visits to members of the medical profession. We shall look at cases in which the art of demonstration depends on professional skills and at others in which the art is treated as a trade in its own right. We shall pay special attention to demonstrator-vendors operating in venues such as supermarkets, fairs, and open markets, or even working door to door.

LEARNING THE ART OF DEMONSTRATION

How does one learn the art of public demonstration? This query can be broken down into several more specific questions. To what extent does the learning process take place through modeling, and, when it does, on what models is it based? To what extent is the art explicitly taught, and in what way? Finally, what skills are transmissible through modeling or teaching, and what room is left for individual initiative and innovation?

To address these questions, we shall look at the contexts and processes involved in learning certain forms of demonstration, demos in particular. Let us start by returning to the investigations I conducted with researchers in artificial intelligence in Silicon Valley at the turn of the twenty-first century, and analyze the way individual actors acquired demonstrative skills (Rosental 2007).

First of all, this apprenticeship appeared to be in part a matter of self-teaching. The skills necessary to perform demos were not taught in college courses. The demonstrators nevertheless had relevant professional skills (e.g., drafting proofs for publication) and various personal experiences (watching film trailers and advertising clips, for example) that could serve as inspiration for forging such skills themselves. Their "training" seemed to be ongoing and reinforced by the practice of demos itself.

Imitation, or modeling, appeared as another important dimension of this apprenticeship. The researchers could take demos produced by their colleagues as examples, or those of major figures such as Steve Jobs. Certain demos had become famous in the world of computer technology: indeed, a demo by Douglas Engelbart has sometimes been called "the mother of all demos."[1] Delivered during a major conference in San Francisco in 1968, that demo presented a computing device endowed with functional capabilities that have oriented the way personal computers are used ever since. This type of demo offered a number of reference points for actors who had to produce their own demonstrations.

Another vector of this apprenticeship was transmission of the art within a "guild" model. While researchers who lived through the "beginning" stages of computer technology early in their careers could count on no one but themselves, for later generations the art of the demo could be transmitted in the form of advice and encouragement from experienced colleagues. In this context, I observed the presence of "chief demonstrators," a term borrowed from the high-tech industry, where it characterizes individuals who are particularly skilled in the art of demonstration. In the industrial sector, in fact, the term can be a job title. A "chief demonstration officer" is generally a specialized professional responsible for carrying out frequent product demonstrations within an organization; as we have seen, Jim Grubb has had this position at Cisco.[2]

The chief demonstrators whose practices I was able to study proved capable of giving lessons in demonstration. They were generally the first to incite their colleagues to present their projects in the form of success stories, and to advise them about how to achieve that goal. A concern for demonstrations was already manifest in the recruitment criteria in the research institutions involved, moreover. The qualities required for producing and performing demonstrations were among the aptitudes sought in the hiring process. The young recruits who manifested these qualities were thus equipped to develop their skills further under the guidance of chief demonstrators.

Good scientists and good managers endowed with good social skills and significant social capital, by and large, these chief demonstrators were able to determine whether demonstrative scenarios proposed were well adapted to the target audiences, and they could suggest how to finetune as appropriate. For although these demos were open to multiple interpretations (that is, the same demo could be grasped very differently from one spectator to another), the fact remained that adapting them to particular audiences in order to maximize their effectiveness was a significant dimension of the demonstrator's work.

Familiarity with the institutions involved, knowing who the interlocutors were, awareness of relevant circumstantial factors and of possible diplomatic biases, mastery of appropriate procedures for preparing the "terrain," understanding what kinds of propositions were apt to seduce or infuriate the audience—all these were precious resources for ensuring the success of demonstrative operations. Having acquired such knowledge, the chief demonstrators could transmit it to less experienced colleagues and orchestrate demonstrative efforts. Access to this shared knowledge also explains why demonstrations prepared by members of the same team rarely stood out as unique, even if the register of the spectacular might suggest otherwise. To the contrary, the similarities between demo scripts and commentaries were quite noticeable: they were grounded in the collective dynamics that governed their development.

The framing contributed by chief demonstrators extended to other forms of demonstration, such as, for example, the sequence of arguments included in research projects destined for select audiences. Here again, the chief demonstrators encouraged their colleagues to demonstrate that their promises were credible and would lead to success. The following commentary, addressed by a manager in an email to one of the researchers who reported to him and had submitted a draft of a project, illustrates this dynamic:

> I'm a little concerned about the overall dark tone of the piece. Any optimism that one might have after learning of [this] success is difficult to retain given the ever more serious deficiencies that are revealed one after another. What it needs is more expressed optimism about our ability to fix these problems. After exposing a problem, you already suggest a direction that will need to be pursued; you just don't indicate why we think that we can accomplish it in the near term. Where possible, I think that you should indicate more strongly that we know what needs to be done, we know how to do it, that it will not take years of research, and any experience that we have that backs up these claims.

In this passage, we see how a chief demonstrator freely dispenses advice to a colleague, indicating in particular how the demonstration could be reinforced to make its promises more credible. It is clear how this type of skill could be transmitted in the context of encouragement within a hierarchy. Acquiring demonstrative skills thus appeared to be a matter of continuing education over a good part of a researcher's professional life.

The interview I had with a French researcher in artificial intelligence in which he described certain aspects of his own trajectory highlights a similar dynamic. This director of an artificial intelligence laboratory explained how he had gradually realized the importance of demos for attracting American and Japanese industry leaders and executives, who had been found to be less receptive to published materials. To that end, he had hired an engineer to construct adapted prototypes. It had taken many months, moreover, to perfect satisfactory repertoires of commentary that presented these prototypes in action (Rosental 2008).

Conversely, several years ago a French researcher in fuzzy logic working in the private sector shared with me an experience that he considered characteristic of the excessive fondness for mathematics—excessive because industrially disastrous—of people he labeled French "technocrats." After trying without success to convince the research director of a major French industrial group of the value of fuzzy logic by describing the principle behind the working of certain devices, he changed his approach and evoked a formal result related to that theory. It was at that point that he discovered the merits of the theoretical register, which won over his interlocutor where his demo had not succeeded. As part of his ongoing self-education, then, this demonstrator had realized that different forms of demonstration were required according to the situation and the interlocutor.

Determining the content of demonstrative registers can similarly be the object of a gradual apprenticeship, as I have already suggested. This phenomenon appeared clearly in an interview I had several years ago with another French researcher in fuzzy logic. The latter explained how he had faced a number of criticisms bearing on his field of research, and how over time he had developed counterdemonstrative repertoires. He had gradually identified certain "standard" critiques, to the point where he could constitute what he called a "blooper" collection. Confronting these recurring critiques, he perfected his own standard demonstrations by testing them in various circumstances until he could consider himself "bulletproof." He had also trained himself in dialectical modes of exchange adapted to his interlocutors, who came from a variety of disciplinary horizons. He was thus confident that he could offer counterdemonstrations in any and all circumstances, either orally or in writing.

In the realm of science and technology, another resource for learning the skills useful for performing public demonstrations, demos in particular, came from advice proffered by various communications consultants such as Carmine Gallo. As we have seen, Gallo's recommendations were based on Steve Jobs's practices, and they echoed the skills and knowledge of various other technology evangelists.

Advice like Gallo's was addressed not so much to researchers in artificial intelligence, however, as to start-up developers and various leaders in the high-tech industry who sought to make their own presentations of products, especially in the form of demos. The lessons of consultants like Gallo took the form of works comparable to recipe books. They presented tips of all sorts for achieving "good" demos. Moreover, these consultants offered apprentice demonstrators individualized, interactive coaching. We have seen how the technology evangelist Guy Kawasaki lavished advice via the internet about demo production in the context of a meeting he was organizing. During that meeting, start-up founders were to perform demos of their products before an audience of investors and journalists.

The lessons of these specialists bore on the multiple skills deemed necessary for the practice of demos. Let us recall that some of these skills had to do with the development of scenarios, others with the preparation of performances (e.g., settings, costumes, lighting, equipment verification, rehearsals), while others applied to the performances themselves. In the latter respect, the advice dealt bore in particular on managing unexpected incidents, on oral delivery and gestures, and on the actor's self-presentation more generally. These recommendations give us a sense that demonstrators had to acquire the skills of a screenwriter, an advertiser, and a salesperson, and also those of an orator, an actor, a director, and a stage manager.

Still other elements enter into play if we include the results of additional studies. When I looked into the background of demonstrator-vendors representing pharmaceutical companies, for example, I learned that a professional magician had trained some of them in techniques of sleight of hand, deemed useful for demonstrations in doctors' offices. In short, the nature of the skills mobilized for the practice of demos and their modes of acquisition and transmission seem both highly diverse and highly complex. This is also true, of course, for other forms of public demonstration.

Let us look, for example, at demonstrations that rely on PowerPoint software. In the world of research, the modes of acquiring the skills needed to perform such demonstrations appear to be located at the intersection between mimetic practices and various disciplinary constraints. These latter nevertheless leave room for the actors to develop individual styles and to contribute collectively to the evolution of the genre (Tardy and Jeanneret 2006; Gaglio 2009; Stark and Paravel 2008).

For a different sort of example, we can turn to street protests in France. For some years now the police officials in charge of overseeing such events have seemed willing to play the role of advisers to inexperienced organizers, provided that the causes being supported strike the officials as legitimate and that there are no objections from the police hierarchy.

Their advice bears primarily on organizational aspects such as logistics and the itinerary of the march (Fillieule 1999).

The acquisition of skills in this area may occur via other paths, however. In France, union organizations in particular can be important contexts for training street protest organizers. Union members can learn to manage various logistical aspects, including communication strategies and police and media relations. They can learn how internal security services work, how to prepare and distribute print materials, and how to organize a march in spatial and temporal terms, even how to formulate slogans.[3]

The existence of a such a framework for apprenticeship thus offers resources for understanding why we find routine, even canonical elements in street protests, and why in that context we can find relatively stabilized genres (Cefaï 2007, 500). Moreover, while organizing street protests is certainly not the defining activity of a given profession in the context of its union organization, any more than it is the defining activity of the police officials charged with managing such protests, it does require a set of skills that can be acquired through practice and transmitted by experienced individuals.

This example raises a more general question: in what cases does the term "demonstrator" refer to the mastery of professional competencies and in what cases does it designate a trade in its own right? Let us look into this question after considering some definitions.

PROFESSIONAL COMPETENCIES AND DEFINITIONS OF TRADES

Dictionary definitions of the term "demonstrator" portray the individuals in question in more than one way. Depending on the context, this category refers either to a specific trade or to an occasional activity exercised in various professions. For example, after recalling that the term comes from the Latin *demonstrator*, "he who shows, demonstrates, describes," the *Dictionnaire de l'Académie française* proposes two distinct meanings for this word.[4] In the first, "older" sense, it designates a "person who teaches or exhibits while showing the things of which s/he is speaking (especially, [a] person charged with making a scientific demonstration before an audience)." In a second sense, it refers to a "person whose trade is to explain to potential users how a device works, how to use a product." In this latter sense the dictionary offers two examples: a demonstrator at a car show and a demonstrator in a beauty salon.

Even if this definition does not give an exhaustive account of the diversity of demonstrators' practices and the contexts in which these practices are deployed, it presents a dichotomy that is worth analyzing. It suggests in effect that the activities of demonstrators in a commercial context (e.g., a demonstrator-vendor or a demonstrator in a showcase of some sort) correspond to those of a trade, whether that trade is exercised temporarily or on a regular basis. By contrast, the posture of demonstrator adopted in the context of classroom instruction or scientific presentations is not that of a specialist in demonstration but that of an occasional practitioner. We can test this distinction by looking at several cases.

In the realm of science and technology, numerous observations show that the posture of demonstrator indeed corresponds to an occasional activity. In my work with researchers in artificial intelligence in the United States, it became clear to me that producing demos, publishing written demonstrations, and communicating them orally were important activities in their field. Not only were these researchers subject to the academic imperative "publish or perish," they also were expected in many instances to conform to the slogan "demo or die." Evidence of this pressure can be found in their CVs. For instance, one member of the Delta project indicated that he had "contributed to writing articles, carried out demonstrations, and given talks." Mastery of demo preparation and performance was valued on the same basis as that of written or oral presentations.

The role of demonstrator is thus by no means reserved for celebrated technology evangelists. Researchers in artificial intelligence, like many other computer technologists, consultants, marketing agents, and entrepreneurs of all sorts, have to step into that role more or less frequently. In that context, the imperative to emulate may extend to the attitudes and even the style of dress adopted by such "visionaries" as Steve Jobs. For example, demonstrators making presentations at Microsoft's 2015 TechDays often adopted iconic elements of Jobs's clothing, wearing jeans and black pullovers in particular, or signaling their casual approach by appearing in T-shirts even before large audiences. This style of dress became a costume, even a uniform, that helped establish the figure of the demonstrator in the high-tech arena. Demonstrators who dressed this way manifested the "cool" attitude with which Jobs made his own presentations. If their behavior at TechDays differed in some respects from that of their model (e.g., they displayed less exaggeration and more "authenticity," a "softer" tone, touches of slightly off-kilter humor, more pronounced irony and self-mockery, a slight detachment), they nevertheless remained in sync through their attire.

The more or less frequent adoption of a demonstrator's posture in the world of science and technology does not necessarily preclude the adoption of other attitudes, however. The researchers in artificial intelligence I met were by no means full-time demonstrators. As we have seen, in this field those who produced public demonstrations were also consumers of the same type of "product," for they themselves attended demonstrations given by their peers. In this context, there was not a stable separation between a group of demonstrators on the one hand and a group of demonstratees on the other. The situation did not imply a stratification in which a dominant group reproduced itself in contradistinction to the members of a dominated group. Each participant was led to adopt the roles of demonstrator and demonstratee more or less in alternation.

To be sure, this picture has to be nuanced within the broader world of research. For instance, mathematicians are major producers and consumers of demonstrations in the form of written and oral presentations. Demonstrative activity thus appears to be a significant component of their trade. But it appears that the more advanced mathematicians are in their careers and the more recognition they have achieved, the more they tend to focus on developing conjectures, leaving to their younger colleagues the task of developing the corresponding demonstrations rather than producing their own (Rosental 2008). Accordingly, the role of demonstrator can be subjected to a logic of stratification. The context of training in anatomy and physiology provides another example. At many points in history, demonstrators have served as scientists' assistants (Hankins and Silverman 1995). The scientist would expose the theory, while the demonstrator would carry out manipulations and show the audience what they should look at. The demonstrator's status was decidedly inferior to that of the scientist.

In the realm of contemporary science and technology, moreover, the practice of demonstration can become a trade in its own right. This is quite clear for example in the case of Jim Grubb, Cisco's "Chief Demonstration Officer," or Guy Kawasaki, former "Chief Evangelist" at Apple. The activity associated with such positions consists first and foremost in preparing and carrying out demos of the company's products. Individuals with such titles perform demonstrations without having been involved in the development of the technologies in question, unlike founders of start-ups, for example. "Professional" demonstrators of this sort may give dozens of demonstrations a year, especially in the context of new product launchings (Markoff 1996). Their skills in this area allow them to pursue careers in the sector of marketing technologies.

Such a trajectory may take demonstrators beyond the company they had represented initially and allow them to become communication consultants, authors of specialized books on marketing techniques, or journalists in the field of new technologies. This type of evolution tends to correlate with a charismatic profile and with talents like those of well-known televangelists or rock stars, allowing professional demonstrators to address audiences numbering in the hundreds, possibly receiving standing ovations. Their skills may lead these high-profile demonstrators to make "brands" of their own names, each taking on an iconic image with a characteristic way of dressing, and each profiting from personal reputation and celebrity.

Of course, the contemporary period is not the first in which the notion of demonstrator has exceeded dictionary definitions or analytic categories and has referred to a trade in the realm of science and technology. Throughout history, the label "demonstrator" has been associated on occasion with a professional status. It was used to characterize the author of lessons in botany in a royal garden, for example, or for the author of anatomy lessons (Hankins and Silverman 1995; Van Dijck 2001). In the pharmaceutical industry, the exercise of demonstrations also became a trade in the United States between the early 1940s and the late 1950s, even if it was generally known under a different name. This case warrants a closer look.

Historical investigations have made it clear how demonstrator-representatives who practiced "detailing" in a particular type of commerce became legitimate professionals in the health system. "Detailing" referred "to the unique performance, half sales pitch and half educational service, with which pharmaceutical sales representatives present physicians with prescribing information, or 'details,' concerning new medications."[5] From 1940 onward, most of the major pharmaceutical companies in the United States had their own "detail men." A massive recruitment effort took place among graduates of pharmacy schools, who were almost exclusively white men. In the late 1920s there were about two thousand such men; by 1959 there were fifteen thousand. This increase paralleled the appearance of a large number of new medications on the market, and a striking increase in sales of such products: from $300 million in 1939 to $2.3 billion in 1959.

To gain acceptance among doctors, the detail men tried to shed the image of specialized salesmen. They sought to appear as competent, respectable professionals playing a vital role in disseminating scientific information. This was not an easy task. Their social status and their income were lower than those of doctors, and they were often less than welcome in doctors' offices. Thus they had to develop various strategies for acceptance. Some of these strategies had to do with appearance and behavior, as can be seen by studying their training manuals, which proposed codes of ethics and precise codes governing interactions with doctors.

One manual spelled out the essential qualities of a detail man: a smart appearance, an engaging personality, a pleasant voice, moderation in oral presentation, good health (not overweight), an alert, agreeable, but forceful manner, and integrity. The demonstrator had to be someone who paid his bills on time, had a cooperative wife, and had a good education adapted to the trade. He also had to keep up with scientific advances, have a solid background in sales, and be capable of working with or leading a team.

Another manual set forth the requisite qualities for being a good detail man in somewhat similar terms: competency, prudence, clear diction, moderation, impartiality, good manners, and a sense of enterprise, fairness, and loyalty. In addition, the detail man needed a good memory and an impeccable appearance; he needed to be observant, patient, and reliable. He had to have a sense of humor and a strong voice; he must not be overweight.

Team managers watched over the appearance and conduct of their detail men. To stay within the norms, the men had to follow a healthy diet, engage in physical exercise, and drink only in moderation. They had to have good posture, be polite and well mannered, and respect social conventions. Their clothing had to be neither too casual nor too sophisticated. The rare women in these jobs had to be attractive but not "too" beautiful. During the hiring process, some companies even conducted interviews with candidates' wives, in order to verify that these women could tolerate long absences on their husbands' part. Detail men indeed had to spend long stretches of time on the road in order to call on as many doctors as possible in their widespread territories.

In order to be perceived by doctors as respectable professionals, detail men had to develop a variety of strategies directly relating to the way they carried out their demonstrations. Having to transmit educational messages about new therapeutic treatments to doctors who were supposed to be in positions of authority on such matters placed the detail men in a delicate situation. They had the difficult task of avoiding giving the impression to doctors that they were being told how to do their jobs even while leading them to modify their practices.

One strategy was to act as if the doctor was already familiar with the products and the results to be presented. To this end, the demonstrators drew on formulas such as "I imagine you're already familiar with . . ."; in contrast, they had to banish phrases such as "I'd like to teach you . . ." Another strategy to avoid upsetting the relations of authority was to reproduce the words of other doctors, citing publications by certain doctors or conversations with them. The manuals stressed that the detail men should be careful to avoid arguing about what was appropriate or not from a therapeutic standpoint; instead, they were to give the impression that they were doing the doctors a service by providing information. This stance helped them avoid coming across first and foremost as salesmen.

Managing relations of authority was especially difficult for beginning demonstrators. Many experienced a certain apprehension at the idea of interacting with doctors. To overcome their fears, they were encouraged to use specific tactics. First, they needed to be very familiar with their products. They should view doctors as ordinary men while treating them with great respect; they should wait for their interlocutor to be seated, for example, before sitting down themselves. They should try to sit near the corner of the desk rather than straight across from the doctor; the idea was to be seen as an ally rather than as a patient, but not as a colleague sitting alongside the doctor. Similarly, to conquer his own apprehension and demonstrate respect for the doctor while appearing as a partner, the detail man should adapt the strength of his handshake to that of the doctor's, should not set his hat down on the desk, and should not smoke. The latter precaution also allowed the detail man to keep his hands free for the demonstration.

While detail men as individuals may have met with mixed success, on the whole their strategies helped to establish the presence of professional demonstrator-vendors in the institutional landscape of US medicine. However, detail men are not the only workers in the tertiary sector who have served as demonstrators or at least used demonstrative skills in the exercise of their profession. We find another example in the case of lawyers, who have to develop demonstrations and present them in public in the form of legal arguments. Generally speaking, lawyers tend to carry out multiple tasks in the context of their work. Demonstrative activity is practiced in variable proportions depending on specialization (criminal lawyers make pleas more often than lawyers in other branches, for example). While this activity may be marginal for some, it is a crucial professional competency for others.

In a very different realm, that of film production in India, the capacity to produce demonstrations is also an essential ability for some of the trades that come together on a set (Grimaud 2007). Film scenarios are generally rough outlines, and filmmakers often rely on assistants to fine-tune the dance and action scenes that structure Bollywood productions. These specialists imagine the details, for the most part, and show the actors and dancers what moves they are to make. Here we are dealing with authentic demos, to the extent that the assistants explain how a stunt or a choreographic sequence can "work." Oral instructions are supplemented by gestures intended to demonstrate their feasibility. Once the movements and actions have been modeled, the actors can work on producing them.

Demonstrations thus constitute an important dimension of the role of specialized assistant directors. The demonstrators themselves often participate in films alongside the stuntmen or dancers they have coached. So here we have dancer-choreographer demonstrators and stuntmen-assistant director demonstrators, for whom the ability to produce demonstrations constitutes a major professional skill.

In yet another sector, that of advertising and communications agencies in France, demonstrations realized with the help of PowerPoint presentations similarly appear to be constitutive components of a professional skill set (Yates and Orlikowski 2007; Tardy and Jeanneret 2006; Gaglio 2009). They entail skills that consultants have to master in order to appear as competent individuals, and they help define professional identities.

Many executives in various sectors of activity also carry out public demonstrations via PowerPoint and attend those of colleagues in the context of meetings or trainings. Even if those practices do not define trades on their own, they constitute significant professional competencies. Michel Houellebecq's *Whatever* (2011) offers an illustration: it depicts in fictional form a depressed French information technologist whose job entails training people working in a government ministry.

The context of sales is another important one in which an aptitude for producing public demonstrations can represent a fundamental professional skill, or even the core of a trade. We have already seen examples in the cases of detail men and Hungarian bankers. But these specific cases scarcely hint at the scope of the phenomenon. Demonstration sales, whether in the form of door-to-door visits or operations in stores or supermarkets, constitute a massive set that merits in-depth examination.

DEMONSTRATOR-VENDORS

In order to carry out an initial exploration of the positions associated with the function of demonstrator-vendor, I analyzed a set of job offers corresponding to that activity that were published in France and in the United States between 2010 and 2017. I also studied a series of job profiles describing the expectations related to this type of employment. These profiles, published on various internet sites, were posted by both private and public institutions.[6] My examination of these materials made it possible to identify several profiles of demonstrator-vendors; I propose to begin with the one that usually corresponds to the labels "commercial demonstrator-animator" and "demonstrator-brand ambassador."

According to the data I was able to gather, the work of these demonstrators consists generally in promoting and selling one or several products of a specific brand over a short period of time, in the context of a promotional event. It may involve, for example, showing how an electronic device works, or offering a food sample to taste. The activity may have several goals, such as introducing a product, spurring sales, and obtaining feedback from the experience of potential or actual clients. A related goal may be to create an event at a commercial site in order to help attract visitors and ensure client loyalty.[7]

This type of demonstration can be used in stores of any size, in showcases or salons for a product sector (e.g., the Salon de l'agriculture, the leading agriculture showcase in France, held annually in Paris), at fairs, or during events of all sorts (e.g., product launchings and Christmas markets). The demonstrator-vendor is sometimes characterized as a "brand ambassador" when he or she adopts that role for the company represented. Such a demonstrator may work alone or as part of a team reporting to a head demonstrator-vendor.

This work seems to be carried out most often in the context of specific missions that are scheduled for a weekend, a holiday, or for an event of several days' duration.[8] But the mission may last longer; it may be spread over several weeks. The demonstrator may be called to work at the stand of a product brand in a department store,[9] or to do a series of sales demonstrations for the brand at various commercial locations (working perhaps sixteen hours a week over a two-week period,[10] or even several weeks in a row[11]). The demonstrator-vendor may be employed by an agency specializing in temporary employment, hostessing, or marketing for a particular company, or by the company itself.

In France, this kind of job seems to be addressed especially to secondary or postsecondary students who are enrolled in programs related to sales and marketing. The jobs are often poorly remunerated internships or short-term contracts paid at a level close to the minimum wage. Some more stable positions of this sort are offered by companies to individuals who have completed the level of education mentioned earlier. But these cases do not always involve organizing sales events.

In the United States, large numbers of demonstrator-vendors exercise the activity of animator or brand ambassador by working part-time for major big-box stores. They may have relatively stable employment obtained via specialized agencies (e.g., Crossmark or Club Demonstration Services through the Costco hypermarket chain). Their work often consists in promoting products of different brands. This type of job tends to be filled above all by individuals with few qualifications and by older people seeking means of subsistence or supplementary income. They must be ready to accept low salaries and flexible working hours, and to work on weekends and holidays. A high school diploma and some experience in sales, as a product demonstrator, or as a brand ambassador, or experience in customer service or in the restaurant business are considered assets without being prerequisites.[12]

Employers seem to expect demonstrators to be capable of attracting clients to their stands or to their areas in a store so these shoppers will take an interest in their products, discover how they work or how they taste, then proceed to make a purchase.[13] Job offers are often explicit on this point. Here, for example, is the profile of a demonstrator sought by the Crossmark company to energize the sales outlets of a supermarket chain in the United States:

> Product / Event demonstrator. Part time. . . . Our Event Specialist team members are brand ambassadors to our clients. . . . Responsibilities: Create a positive product impression and generate brand awareness for onsite/in-store food and non-food demonstrations. Proactively intercept, engage, interact and communicate with consumers to demonstrate products features and benefits in a professional manner. Possess the ability to acquire and maintain knowledge of products represented. Properly prepare event table/cart (set up & breakdown) for demonstration execution. Ability to prepare and serve food and beverage samples using required appliances, such as: microwave, convection oven, electric fryer, skillet, juicer, coffee maker, cooking utensils, forks/knives and hot oils. Maintain a clean, sterile and safe work station using cleaning chemicals. Maintain an overall professional appearance consistent with the requirements of the role. Provide excellent customer service and develop a professional working relationship with store personnel to effectively meet Crossmark's and client

objectives. Accurately prepare and submit all online requirements on the day of event demonstration/execution.

This listing brings to the fore a certain number of aptitudes that individuals seeking a job as demonstrator at Crossmark should possess. In particular, they must be able to interact with supermarket clients, promote products, manipulate various devices, manage their stand, and also carry out tasks such as reporting and cooperating with other members of their professional milieu. A job listing for a demonstrator in France for a major television brand conveys similar expectations: "You are the ambassador for a world-famous television brand, within a specialized store like Darty or Boulanger. As such, you are responsible for demonstrations, client counselling, and sales of TV products, interacting with the clients of this store for the purpose of increasing sales. . . . You are to make weekly reports on your activity via a specialized tool."[14]

Similar qualities are also often sought for more itinerant positions, as we can see from another listing published by a French agency seeking a demonstrator-vendor who can go from store to store to promote the products of a company that makes alcoholic drinks. The job offer in question specified that the candidate had to be able to "call attention to the client's products in the alcoholic drinks section [of various stores], advise consumers, ensure the promotion of the brands, and run tasting events."[15]

This type of listing concerns the relatively stable positions of demonstrators. But, as we have seen, a number of job listings are for temporary postings or internships. This is how a manufacturer of window shutters sought a demonstrator-vendor intern to work one summer at his stand in a Paris department store. Candidates had to be able to ensure "product staging," "welcome, advise, and win over clients" and "blow up the bottom line."[16]

In another listing, a French agency said it was looking, on behalf of an eyewear brand, for "demonstrators/sales event organizers" to work for two- to four-week periods during the summer; candidates had to be "dynamic and cheerful" and have "an excellent feel for human relations, for business, and for hosting."[17] Similarly, an agency specialized in event planning in France sought, for a one-day promotional operation, "commercial event organizers to promote a famous brand of Belgian chocolate . . . host tastings of these famous chocolates in supermarkets so as to increase awareness of the brand . . . a report on the event would be required at the end of the mission."[18]

Finally, we can look at the case of another French agency specializing in event planning that sought, "in the context of a commercial operation . . . involving all the distributors of a footwear brand . . . two demonstrator-trainers . . . tasked with training the staff at the points of sale and then organizing the commercial operation on site."[19] In this case the tasks of demonstrating and organizing were combined with that of training sales personnel.

These listings and others like them presented various skills and qualities as essential for holding this type of job in France. If we aggregate them and review the categories generally included, we observe that demonstrators must be dynamic, enthusiastic, smiling, tenacious, serious, rigorous, organized, autonomous, available, endowed with a good sense of interpersonal relations and with a knack for hosting, for selling, and for listening to clients.[20] Experience in sales and in organizing sales events is sometimes also sought, along with a certain mobility, for positions requiring travel. Evidence of aptitude for teamwork may also be required when the candidate will be joining an organized group of demonstrators.[21]

French job profile postings also insist on the need for demonstrators to know their products very well, so they will be able to "explain in what respects the product is of high quality, innovative, practical, or better than another." Demonstrators must also possess various traits (for instance, poise, spontaneity, conviviality, a sense of humor) that are thought to allow them "to get closer to their interlocutors," an achievement that may "weigh in the final decision to purchase" the product being promoted.[22]

In the United States, job listings that are more detailed, often for legal reasons, stress additional skills and qualities as required for employment as a demonstrator. These may include mastery of digital tools that allow employees to access various databases and training programs as well as to produce reports. Basic physical skills may be required, and a capacity to follow instructions relating to appearance and dress. The following excerpt from a listing posted by the Crossmark agency (which hires demonstrator-vendors on behalf of big-box chain stores) is illustrative:

Physical Appearance. Associates must present a professional, neat, clean, and appropriately groomed appearance. Specific Retailer Dress code requirements are: long-sleeve, button-down white collared shirt along with full length black pants, and solid black, low-heel, closed-toe/closed-heel shoes. CROSSMARK will supply a baseball type cap, and an apron.

Health Code Requirements. Hair must be tucked into the ball cap or hair net, exposed facial hair may require a beard guard. No facial jewelry, including tongue eyebrow or nose piercings. Tattoos which are not covered by the uniform including face and neck must be covered with a bandage or makeup . . .

Physical Demands. The employee will be regularly required to: Stand up to 4 hours at a time without a break; Walk; Use hands and fingers to handle or feel; Reach with hands and arms (including reaching overhead); Talk and hear; Visual ability to read instructions and perform events; Stoop; Kneel; Crouch; Climb (including use of a 6' ladder); Balance; Lift and carry up to 25 pounds (including occasional lifting of up to 50 pounds); Push and pull a wheeled demonstration cart weighing up to 300 pounds fully laden with appliances, supplies and product; Prepare foods and beverages using the required appliances, such as cooking utensils, knives, convection oven, fryer, coffee maker, electric fryer, microwaves and hot oils; Be in contact with cleaning supplies. Orange Juicer: some demos require routinely lifting of up to 50 pounds. Pineapple Corer: equipment contains sharp edges. Tortilla Maker: exposure to heated and pressurized equipment.

These descriptions emphasize the fact that the job of demonstrator-vendor may entail not only "representation" of a brand or product, but also physical tasks that go well beyond the use of speech, potentially engaging the entire body. These tasks include various manipulations and movements such as lifting and carrying heavy weights or remaining on one's feet for up to four hours at a time. Jobs such as these seem to belong to both the manual labor and service sector categories.

Not all demonstrators function as event organizers and sales representatives in commercial contexts, of course. Others practice their trade by calling on professionals, advising them from a distance, providing training services, or a combination of these. The positions held by the demonstrators in this group require higher qualifications and tend to be stable and better paid. They are reserved for individuals with experience in a specific sector of activity and often with some higher education (e.g., at least two years beyond high school).[23] The detail men working in the pharmaceutical industry are good examples of demonstrators in this category. But a study of job listings reveals many other cases corresponding to this profile.

Let us look first at a listing posted by a manufacturer of intraoral digital scanners seeking a demonstrator-trainer in France. Candidates must be able to travel in order to call on dentists in their offices, show them how to use the products, and offer them follow-up assistance, either on site or remotely.[24] In another example, a French employment agency indicated, on behalf of a producer of software packages, the need for an individual who could fill a position in "client relations as a pre-sale demonstrator." This individual had to be able to carry out "multiple tasks requiring autonomy and responsibility" within a team, to intervene "between the commercial party and the technical party" and constitute "a strong link in the sales process." The candidate had to be able to "carry out presentations . . . by telephone or videoconference; advise, listen to, and analyze clients in order to give constructive feedback to the commercial party; train clients and provide follow-up service."[25]

In another job listing, a company in the field of "commercial and technical deployment of fiber-optic networks and telecom solutions" sought a demonstrator-vendor in France with "successful sales experience" to take over "sales of [its] products and services" and to develop and ensure the "loyalty" of its "client portfolio." This position was connected with a team and could "evolve toward management positions."[26]

In the context of such listings, the demonstrators the companies are seeking appear to be individuals capable of performing demos in order to carry out trainings, provide technical assistance, and make sales. Their demonstrative activity should allow them to collect data on their prospects so they can propose targeted products, in liaison with a company's marketing and technical services. Various specific qualities and skills are required for this type of job, such as mastery of specific digital tools, skills in communication and "interpersonal relations," and a talent for training and marketing. Demonstrators must also have an aptitude for listening and for negotiating, coupled with skill at getting their clients to modify their own practices.

The terms used to characterize the individuals sought for such positions is noticeably different from the ones that characterized demonstrator-organizers or sales representatives. The new list includes the following descriptors: positive, enthusiastic, autonomous, responsible, organized, punctual, experienced in the sector, apt to take initiative, prepared to travel frequently, and capable of working by telephone and as part of a team.[27] These qualities and skills are in harmony with the higher status of this type of job as compared to those of the demonstrator-organizers described previously.

After examining these diverse demonstrator profiles, we have to recognize wide variations in the nature of the jobs in question, along with significant phenomena of stratification. The status of a demonstrator varies according to the qualifiers appended to the label (e.g., vendor, organizer, ambassador, trainer). At one level, we encounter jobs with few qualifications, marked by certain physical demands and considerable variability in work hours. These jobs, at the bottom of the pay scale, go mostly to temporary or part-time workers and to interns. At a higher level, we find more stable and better remunerated jobs that require more highly qualified candidates. These positions, designed for full-time employees and specialists in technology sales, for example, are in turn distinguished from those of "demo gods" such as Kawasaki and Grubb.

It is clear that we need to speak of jobs as "demonstrator" or "demonstrator-vendor" in the plural and as multifaceted. Their diversity is all the more pronounced when we add demonstrator-vendors who work at fairs and in marketplaces to the mix. The profiles of these latter differ from those we have already examined.

The trade of demonstrator at fairs and markets is an ancient one (Duval 1981). In France, it corresponds at present to a particular type of non-sedentary vendor. In French administrative parlance, this category refers to a status that generally permits such persons to manage a certain number of sites in marketplaces.

A typical set of regulations governing public markets, issued by municipal authorities and published in a monthly newsletter for non-sedentary vendors, refers to four types of sites. "1. Those reserved for subscribers (70% maximum). 2. Those reserved for occasional vendors and itinerants (20%). 3. Those reserved for demonstrators (5%). 4. Those reserved for *posticheurs* (5%)." A demonstrator is defined as "a non-sedentary vendor presenting in the public domain (markets, fairs, trade shows, . . . etc.), on an occasional basis, a device or a product whose functions he explains, whose use and benefits he demonstrates, and which he proceeds to sell." This type of trader is distinguished from a *posticheur*, who is described as "a non-sedentary trader presenting on an occasional basis in the public domain (markets, fairs, trade shows, . . . etc.) diverse merchandise sold by lot or by the piece (tableware, tools, household linens, jewelry, or bakeware)."[28]

These definitions thus establish a difference between vendors who carry out manipulations designed to accompany arguments and explanations about the way products work (demonstrators) and those who basically use language alone (posticheurs).[29] The connotations of posticheur, or "smooth talker," are in fact quite similar to the notions of "hawker," "peddler," or "huckster," referring to "one who sells or advertises something in an aggressive, dishonest, or annoying way."[30] However, it does not necessarily convey the same pejorative meanings. It may refer in a somewhat more neutral way to a trader who uses some oratorical talent to attract people and sell them some goods.[31]

The foregoing definitions stress the fact that demonstrators and posticheurs (categories 3 and 4) are not present on a regular basis at a specific marketplace (unlike "subscribers," category 1), and that they are in fact itinerant vendors. Nevertheless, they are distinguished by their sales methods from other "occasional" or "itinerant" vendors (category 2) who are presumed not to practice either demonstration or hucksterism.

The regulations I have just cited identify other types of non-sedentary vendors who may turn up at markets in France, alongside demonstrators and posticheurs: artisans, producers, vendors of manufactured products or food items, and second-hand or overstock dealers. These distinctions bring to light various activities that are not practiced by demonstrators, as for example the sale of home-grown food products, or garage sales.

The specificity recognized through direct and indirect definitions confers particular rights on demonstrators. The regulations spell these out as follows: "Given the sales techniques both of demonstrators and posticheurs who present articles to larger audiences than those of other vendors, it is appropriate to define commercially viable sites proper to the exercise of these two activities and separated from one another in order to spread the hustle and bustle. As demonstrators are non-sedentary occasional vendors, several commercially viable places of sufficient surface area (4 linear meters) are reserved for them."

This provision brings to light the specificity of the audiences for demonstrators and posticheurs. Their sales methods generally draw crowds, thus calling for particular measures on the part of municipal authorities for ensuring the fluid circulation of human traffic. In addition, we see that the activity of these vendors is once again viewed in terms of commercial event organizing. The attention paid to the surface area of stands underlines the fact that the sales spaces allocated to demonstrators and posticheurs represent stages on which "shows" are produced. This makes it easier to understand, moreover, why the term "barker" is sometimes used to characterize these actors: they must raise their voices to make themselves heard and attract spectators in open areas.[32]

Given the potential for loud noise levels, we can understand why the regulations provide for distancing between demonstrators and posticheurs. Such a measure promotes a distribution of commotion and entertainment within the marketplace. But it also takes material constraints into account. The words of a demonstrator and a posticheur in close proximity would be hard to grasp by the audience that each of these vendors needs to create and maintain before proceeding to make sales. In addition, the crowds could be destabilized by the attraction of spectators to the neighboring posticheur or demonstrator.

The regulations that I have just evoked also emphasize significant phenomena of competition among the various actors in a marketplace, which imply a need for administrative control. Some studies have shown that subscribers tend to monopolize the best sites, to the envy of demonstrators and posticheurs. In addition, vendors in the latter categories tend to attract large numbers of clients, a fact that can disrupt the sales of subscribers; their margins also generally seem to be higher. This context is the source of jealousies and sometimes violent conflicts (Duval 1981).

The analysis of the space granted to demonstrators in French markets and of the categories used to describe the demonstrators thus supplies elements that make it easier to grasp the nature of the trade they are exercising and the rivalries in which they are involved. The examination of the categories used to describe other types of non-sedentary vendors also helps us grasp their contours. This may help explain, at least in part, why demonstrators subject to the regulations outlined above have their own union, in France.[33]

The outlines of this trade can be spelled out even more precisely by examining still other terminological elements. In France, demonstrators are often considered by the public at large and by certain analysts as embodying a particular type of huckster, a *camelot* (Le Velly 2007, 146–147), that is, an itinerant peddler selling objects of little value in the street.[34] Although this lexical usage is not universal (Duval 1981, 147), it brings to light another dimension of the demonstrator's activity, that is, the sale of objects that are of relatively mediocre quality in relation to their sales price. It thus highlights the capacity of talented demonstrators to realize significant profit margins thanks to their oratorical talents and their clever manipulations. In addition, it is a reminder of the fact that demonstrators do not work in stable businesses that they own themselves.

According to some studies (Duval 1981, 146–147; Le Velly 2007, 150), demonstrators who operate in markets and fairs, like other types of *camelots*, constitute a semi-nomadic population in France. Although they often have home addresses, they generally frequent markets and fairs from March or April until late December. They form a relatively closed, homogeneous group, the majority consisting in actors from the least privileged social classes, like their clientele. They seem nevertheless eager to pursue their trade independently, hoping to make more money than they would as manual laborers (Duval 1981, 146).

The trade seems to be largely transmitted from generation to generation. The tricks of the trade, the products, and the most desirable sites in marketplaces and fairs are not easily accessible for anyone who does not belong to an intergenerational network (Le Velly 2007, 150). The relevant skills and knowledge seem to be transmitted largely by modeling. A certain solidarity is perceptible within these networks, even if interpersonal relations are sometimes ambivalent. This solidarity is stabilized by shared techniques, a common jargon, and a collective code of honor. The latter encourages the actors to refrain from criticizing the merchandise of their peers and to avoid instituting price wars (Le Velly 2007).

Nevertheless, the relation of these vendors to their trade seems to have evolved in France over the years. The younger generation appears less inclined to play the role of public entertainer; their practices seem to be more straightforward and modeled on the more strictly governed practices of demonstrator-vendors in supermarkets (Duval 1981). The following account provides an illustration:

> René is ageless. As he tells it, he has always been a demonstrator. When he began in the late 1960s, his mop of red hair was famous in the markets of eastern Paris. . . . Now, René only works occasionally; in any case, he feels that the trade has changed. "My own master taught me to bury the clients under a verbal deluge. Demonstrators [today] are more discreet. They no longer try to make people laugh; they try mainly to advise their audience. That said, I understand them: the pressure of the bottom line is such that a bad demonstrator doesn't last very long."[35]

This testimony makes clear an evolution in the practice of demonstrators in markets and fairs, articulated with the evolution of the commercial goals that motivate them. According to my own observations, the intention to entertain remains present, even if the means used to that end and the forms of entertainment deployed have been renewed to some extent. In the investigation I conducted at the 2014 Paris Fair, I was able to observe how certain demonstrators relied heavily on humorous remarks and managed to attract a large audience with their verbal talent and a certain clever sleight of hand.[36] Moreover, a number of visitors were primarily out to have a good time, whether through the pleasures of shopping and discovering surprising objects or from the delight of wandering about the stands, becoming immersed in a lively atmosphere, or simply getting away from their daily routines. These observations highlight once again the place occupied by entertainment in the practice of public demonstrations.

As we conclude this chapter we have a better grasp of what is involved in these demonstrations. We first saw how the corresponding skills and knowledge are acquired through apprenticeships that may be based on self-teaching, modeling, or intergenerational transmission. In many cases they stem from the mastery of professional skills or the trade of demonstrating itself. The exploration of the conditions for learning the art of demonstrations and the expectations associated with its practice has thus brought to light certain evolutionary developments and specificities of the skills that are implied in various business sectors. While clarifying what makes demonstrative performances possible, this investigation has also helped bring to light a series of constraints that bear on demonstrators' practices and the relation of these practices to other activities (e.g., sales, event planning, training, or outreach). Finally, we have observed that the groups producing public demonstrations were highly stratified and endowed with variable status (from the most modest to the most prestigious), which also had a major impact on the nature of performances.

In this process, we have noted the importance of recreational approaches in the conception and realization of many public demonstrations. This phenomenon raises the more general question of the relation between public demonstrations and entertainment in its various forms. In what cases and to what extent is this dimension superimposed on or substituted for a utilitarian vocation? In addition, to the extent that the entertainment of crowds can consist in a cultural production, a commercial item, or a form of art, one may wonder whether these descriptors are relevant for describing some or all public demonstrations. We shall consider these questions in chapter 7.

7

ENTERTAINMENT, CULTURAL GOODS, FORMS OF ART

To continue exploring the ways public demonstrations are conceived and carried out, let us now look more closely at those that take the form of entertaining spectacles. We shall consider both contemporary examples (e.g., demonstrations at fairs and showcases, street protests, and demonstrations of skill, athletic prowess, or artistic talent) and historical cases (e.g., demonstrations of hot-air balloons or mermaid mummies). We shall also examine an important recent phenomenon, the development on the internet of forms of entertainment related to watching product demonstrations, especially those targeting adolescents; to this end, we shall focus on examples from the realms of video games and cosmetics.

We shall see how certain demonstrations are conceived or exploited as cultural products and as market goods intended to be consumed on a more or less vast scale. Examples can be found dating back at least to the Renaissance, continuing through the demonstrations made by François Bienvenu and by P. T. Barnum, as well as some that were produced at the Château de Versailles during the reign of Louis XIV and are evoked in demonstrations produced today.

We shall also see how public demonstrations may be conceived as forms of art. Looking at the work of several contemporary artists from around the world will allow us to analyze the nature of such performances and see how they are used to present different perspectives on the world, to entertain, to encourage critical reflection, or to instruct.

ENTERTAINING SPECTACLES

When asked how they perceive their audiences, demonstrators at fairs in France are quick to explain that visitors to a fair are generally looking for entertaining spectacles and distractions, while being open to the possibility of making purchases, in the context of an outing with family or friends (Le Velly 2007). Whether it is clearly articulated or not, the two-fold dimension of these demonstrators' activity (entertaining and selling) is largely determined by this understanding of their public (Sherry 1998).

Thus demonstrators often strive to shift fair-goers from the status of passersby to that of consumers, by offering various sources of enjoyment (e.g., tastings, jokes, and tips for everyday life). These maneuvers draw in clients and make them feel they ought to buy something. During their performances, skilled demonstrators can observe which of the visitors are likely to remain onlookers and which are potential customers. For the pleasure taken by at least some demonstrators in having a platform for "putting on their show" and "working the crowd" is combined with a more widely shared desire among these actors to make quick profits that will allow them more income and more freedom than they could earn in many salaried positions (Duval 1981).

The diverting character of the performances is also an asset, in the middle of a crowded fair, when it comes to attracting journalists to ensure media coverage and promotion of the events or products. These observations make it clear that the "entertainment" dimension of demonstrations remains a standard feature in this sort of gathering.

The model of high-tech conferences, salons, and showcases is similar, from this standpoint, to that of fairs. At Microsoft's 2015 TechDays conference, the playful character of many demos—bringing to mind for some the universe of video games (the demonstration of the Spartan project examined earlier is a case in point)—was appreciated by a large segment of the public, as attested by many visitors via tweets. This observation is just as valid for the demos produced during conference sessions as for those carried out at the stands of the salon.

ENTERTAINMENT, CULTURAL GOODS, FORMS OF ART

The demonstration of the 3D printer described earlier offers another illustration of the ludic character of many demos. The somewhat childlike pleasures usually aroused by such demos were often shared by the demonstrators themselves, although the feeling tended to fade after many repetitions of the performances, especially toward the end of a long, tiring day. This phenomenon could be observed especially at stands staffed by demonstrators whose pay was not dependent on the outcome of their encounters and who were eager to break up their long hours on duty with some enjoyable moments spent among the crowds.

In any case, the 2015 TechDays stands that featured augmented reality, connected objects, and robots drew large numbers—more curious onlookers than clients—owing to particularly entertaining demos, as we can see from a detailed study of one such event, a demonstration of augmented reality software that reproduced the movement of a user's hand with the help of a sensor. Users could thus move virtual cubes around and see the results of their actions on screen.

The demonstrators showed curious onlookers how to use the software, explained the principles behind it, and answered various kinds of questions. In the interactions transcribed (in translation) below, one demonstrator, soon to be joined by two colleagues, presented the programming details (which we shall touch on only lightly here) to a visitor introduced by a hostess.[1]

AUGMENTED REALITY DEMO

HOSTESS: Hello! Would you like to participate in our drawing to win a Surface Pro 3?

VISITOR: Why not? What's your field of activity? Are you developing frameworks or apps?

HOSTESS: No, that's not it, it's really just for entertainment purposes. You can buy the headset directly from the company that makes it. Here, we just bought the headset with the software that comes with it, so everybody can use it. No, what we're developing is what you see back here.... [Here], it's augmented reality. So if you put your hands over the sensor, you can play with the cubes and the balls. [A demonstrator takes over.]

DEMONSTRATOR A: Here, you have to raise your hand. Wait, I didn't really put it facing you, it should have been right in front. But you have to see that you're really turning all over the globe . . .

VISITOR: But how can I hold on . . . ?

DEMONSTRATOR A: Here, it's just a trial run. . . . To simplify, we used some, how can I put it, some default mechanisms. So . . . the 3D game engine . . . controls the collisions on its own. We haven't perfected the mechanism, which means that your hand is represented pretty much everywhere by cylinders that come in and bump the cube, let's say, . . . We should have controlled the shape of the hands better. . . . We didn't have time. This version was just a trial run. We told ourselves that, for Tech-Days, well, we wanted to show something nice, something out of the ordinary. . . . You see your hand? It's there, but actually, we're turning it around. Your hand is in position. And there, move it a little. You see, we're watching it from above. We really have the impression of the hand in all directions, with an interaction that's really interactive augmented reality.

VISITOR: There's just one sensor?

DEMONSTRATOR A: And it calculates the position of the hand. . . . That little sensor is swinging an infrared beam back and forth. . . . We just have the information: I'm at this place, in this position, and so on. . . .

VISITOR: You were aiming for applications?

DEMONSTRATOR A: Actually, . . . we wanted to show off our ability to code in new areas. Because we're working with big data, we're doing really serious stuff, but we can have fun, too. . . . We have commissions from big accounts. . . . There, it's much more serious. But we said to ourselves, . . . to show that we actually have that spirit of innovation. . . . It's true that we did this in three days, and we told ourselves that maybe we could . . . [Laughter.] . . .

DEMONSTRATOR B: Augmented reality is around, there are companies that specialize in it, but . . . by showing examples like this, we make it obvious that we have the capacity to do it, and that it's actually not priced out of reach. . . . There are technologies that cost a real fortune. This kind of product costs ninety-nine euros: ninety-nine euros, a webcam. Just an ordinary webcam . . .

VISITOR: And you couldn't have done a spectacular demo with big data? . . .

DEMONSTRATOR A: We did put on one session, I was the speaker yesterday. . . . To be honest, we hadn't planned on doing TechDays. We signed up at the last minute. We put the thing together in three days. We did all we could to show interactivity. But actually, we're specialists in big data. We know how it works, we know the disadvantages. And we might have been able to set up an interaction with it . . . because, around all that, we also did, well, we also put on a session, that takes time. . . . Then we made a start implementing big data, to show how it works. . . . So it's not good for much, it's a robot, but we got it connected to this thing here, with Bluetooth. . . . What we did was . . . the data brought in by this connected bracelet are recovered on the server side, and then we put them into the big data system . . . and afterward, behind the scenes, we showed how the data could be analyzed . . .

VISITOR: Then you can also suggest possible uses?

DEMONSTRATOR A: . . . In fact, we're able to do a bit of everything, anything at all, I'm tempted to say. Uses? Sure, we could have them, we're really there, we're . . .

VISITOR: You're working on assignment?

DEMONSTRATOR A: Exactly.

VISITOR: But if you let yourselves go . . .

DEMONSTRATOR A: That's it exactly! [Laughter.] You've got it! If they tell us, let yourselves go, run with it, full speed ahead, then . . . wow!

DEMONSTRATOR B: Now, we're doing serious stuff. Afterward . . .

DEMONSTRATOR C: Three days, so if they give us a little more time, we can do something better.

DEMONSTRATOR A: That's right! . . . I can give you my card . . . I don't know if you've signed up to win a Surface Pro. [He turns to a hostess and asks her if the visitor can sign up.]

HOSTESS: Yes, of course. . . . There, you're all set. . . . Bye, have a good day!

The way this interaction unfolded shows how demos presented on certain stands were "attractions" in the literal sense. Their performance allowed the demonstrators to arouse curiosity, to attract visitors on the same basis as a carnival game, and to feature products or, as in the present case, to show off skills. Here, a hostess helped draw visitors to the stand and organized waiting lines. One or more demonstrators then performed demos whose structure and associated commentary were adjusted to the questions and reactions of the demonstratees.

The demo just transcribed had been conceived as a playful activity, both for the visitors and for the demonstrators themselves, as we see from the way the demonstrators contrasted the seriousness of their professional missions with the exercise they had gone through to create their attraction. The impulse to entertain was combined with a desire to impress. It was essentially a matter of highlighting the exceptional polyvalence of the information technologists on the stand and their ability to produce results at low cost and in record time. This two-fold effort to entertain and impress was also apparent in the structure of the other intervention evoked by the demonstrators, in the session that focused on the core of their trade.

The effort to entertain is echoed today in demos produced in a wide variety of contexts, including product launchings, as we have seen. But this is hardly a new phenomenon: similar efforts can be identified in public demonstrations throughout the history of science. Over the centuries, technologies have been the focus of countless spectacles in the form of demos, exhibitions, or publications (Grelon, Chamozzi, and Wagner 2000). For example, fireworks served to demonstrate the power of the aristocracy in European courts from the Renaissance up to the eighteenth century (Werrett 2009). Public demonstrations of "natural philosophy" in England also included an important theatrical dimension and were sometimes used as sources of entertainment (Schaffer 1994).

Public demonstrations of hot-air balloons in the 1780s supply another illustration, as we have seen.[2] These events took the form of spectacles structured in three phases: the preparation for lift-off, the flight, and the return, culminating in the landing. When these spectacles were produced in various cities in France, they drew crowds in which all social classes came together. The events benefited from organizational skills developed in the context of great dynastic festivals and religious processions: real expertise in the staging of spectacles came into play. As in a theater that participates in the stratification of a social order through its pricing structure, at these events there was paid seating in an enclosed space near the action. Entrepreneurs and workers busied themselves on a platform to prepare for lift-off, in the presence of others representing authority and knowledge. Outside the enclosure, however, the spectacle was cost-free.

The various social classes found differing forms of entertainment at such events (Thébaud-Sorger 2009, 214–239). Members of the lower classes and the bourgeoisie could plunge into an atmosphere of festive curiosity as they contemplated unprecedented phenomena. Members of the aristocracy found themselves in the midst of familiar worldly spectacles, experiencing moments of euphoria at the high points, and moments of amusement at the sight of learned men scurrying about; the aristocrats conversed among themselves as they would in their box seats at the theater. The aristocracy's interest in scientific and technological progress was itself visible on stage and became part of the spectacle for the broader audience. The aristocracy in effect presented itself to the population at large as an enlightened elite contributing to the mastery of progress and validating the experimental protocol.

Such public demonstrations thus did more than satisfy the interest of scholars in unusual phenomena (Thébaud-Sorger 2009, 205–242). They were at once ceremonies, events designed to edify, and commercial spectacles; they served as public festivals, diverting events, and proofs of dexterity and bravura. The credibility of the elites was at stake in this form of entertainment. Even as they brought extraordinary phenomena to light, these demonstrations were occasions to celebrate the king as the patron of inventions, along with the local authorities who facilitated the events. They celebrated scientists' knowledge and their power over matter, through the simplicity of the explanations proposed and the mastery of technical procedures. Such demonstrations also provided opportunities for various groups and individuals to promote themselves and their own interests.

Finally, unlike more carefully controlled events such as religious processions, these demonstrations represented a moment of coexistence and interaction among all the components of a society and a city. These festive celebrations of progress gave the entire population an occasion for consensus.

Throughout history, political authorities and organizations have sought to introduce a dimension of entertainment in the public demonstrations they organized, or have managed to draw benefits from those in which they participated. To attract citizens, the organizers of political gatherings such as those of the American Forum Movement[3] in the 1920s and 1930s gave their meetings the form of spectacles (Cossart and Taïeb 2011, 147; Keith 2007). Similarly, many peaceful street protests today have a festive character (Shepard 2009; Champagne 1984). The element of entertainment often represents a resource for mobilizing participants, rallying populations to a cause, and providing materials to journalists that they can use to create a media event (Fillieule 1999; Lemieux 2008).

As we have seen, the Scientists on the March street protest in 2014 was punctuated by dancing, songs, and mini-concerts. Posters, placards, and other banners transmitted messages and drawings that were often humorous. Whistles, musical instruments, and sound systems added to the entertainment, as did costumes and disguises. The audience for this spectacle included journalists, passersby, residents of the neighborhood, and the demonstrators themselves. The static gatherings, the street "sit-ins," and the movement of the advancing marchers fostered moments of conviviality for the participants, providing occasions for colleagues to meet and converse.

This demonstration conveying political demands thus had several features in common with other street spectacles that meld entertainment and demonstration, for example, displays of skill, artistic talent or athletic prowess. I was able to observe an instance of the latter in 2013, at an annual event in Saint-Gély-du-Fesc in southern France: a demonstration of skill by *gardians*, farm workers on horseback who manage herds of bulls in the Camargue region (see figure 7.1).

This event took place in a village street. A number of bulls were released from trucks and guided by horsemen along the route; the gardians on horseback demonstrated their ability to round them up over an area of a few hundred meters and corral them. This exercise, which was not without risk, allowed the gardians to display their prowess and courage before a public consisting of local residents and tourists, crowding in behind sturdy fences.

At the same time, teams from the village competed in a race: they ran behind the bulls, trying to grab a tail and hold on as long as they could. These participants, too, sought to display their skill and courage in a practice that was also risky and exciting. As a competition, it was

7.1 Demonstration of the running of the bulls in Beauvoisin, France, in 2007. © 2007 R. Benezet. Creative Creative Commons Attribution-Share Alike 3.0 Unported license (https://creativecommons.org/licenses/by-sa/3.0/deed.en).

characterized by good humor: it was all in good fun, shared by people well acquainted with the practice.

These demonstrations were at the heart of a popular festival for village residents, and at the same time they were tourist attractions for vacationers present in the region. The running of the bulls, a major annual event for the Saint-Gély-du-Fesc residents, was the prelude to a nocturnal festival enlivened by music. The demonstrations also served to promote the profession of gardian and to support the interests of the *manadiers* (the livestock farmers who raised the bulls), thereby defending a regional tradition. Such demonstrations clearly played many roles at once.

To offer another example, I had the opportunity to observe a demonstration combining athletic and artistic prowess on the Place de la Comédie in Montpellier in 2014. This event featured two artists performing in turn. The first, a contortionist, displayed his extraordinary flexibility by adopting a series of extreme postures. The second explained what a

Human Beatbox was and then gave a demonstration: the performer produced rhythmical music by imitating various instruments, mostly percussive, with his mouth alone.

To the extent that the performances were street spectacles, the demonstrators' first task was to get people walking by to stop and cluster around them. Like demonstrators at fairs who have mastered the art of gathering crowds, they asked individual passersby to approach the improvised stage. Next, they invited certain spectators to move to the side to fill in some empty spaces. Once that first circle formed, it helped attract other curious onlookers. The demonstrators then quickly began to perform, so the audience would not become impatient and drift away

After an initial demonstration of the contortionist's talents, the Human Beatbox displayed other skills echoing those of experienced demonstrators at fairs. He took care, first, to interact with the spectators to make sure they did not leave. In this process, he rapidly punctuated his discourse with questions addressed to them, encouraging their participation. He also practiced storytelling, explaining for example that his parents had not been able to afford the drum set he dreamed about when he was a child. His discovery of the Human Beatbox model had been a revelation.

Following these well-honed preliminaries, the artist began his demo, showing how he succeeded in imitating various instruments. He then performed a few numbers, kept brief to avoid boring the audience. These segments were organized in a sequence that built up to ever more astonishing effects. Once a culminating point had been reached, the two artists announced that they were about to pass the hat, a development that many spectators had not been expecting. This practice could be compared to the culminating moment at a fair when a demonstrator announces the price of the product on display and presses the spectators to make purchases right away.

The Human Beatbox demonstration was thus multidimensional: beginning ostensibly as a pedagogical display, it was also a popular entertainment and a commercial spectacle. Similar operations sometimes add an additional dimension, that of institutional promotion. I was able to observe such an event when the eightieth anniversary of the French Air Force was celebrated with an exhibition of the prowess of the Patrouille de France, a precision aerobatics unit (see figure 7.2).[4]

7.2 Demonstration of aerobatic prowess by the Patrouille de France in 2014. © 2014 Claude Rosental.

The planes in this squadron performed aerial acrobatics above a crowd of spectators that consisted primarily of tourists. Many graphic designs in the sky, magnified by contrails in blue, white, and red, highlighted the pilots' skills. The planes hurtled toward one another at low altitudes, adjusting their flight paths at the last second to avoid colliding. Many spectators, tense and excited, applauded energetically to show their admiration.

This demonstration of skills had a two-fold character. By converting capacities honed for war into acrobatic talents, it became at once a source of entertainment for summer visitors and a promotional operation for the French Air Force. The commercial dimension of the event was marginalized, however; it consisted principally in the sale of various derivative products such as model airplanes and T-shirts from a truck.

Demonstrations of skill do not have to be performed live to be appreciated by the public; many have achieved great success on the internet. We have already seen how demos containing bugs or "bloopers" from video games can draw a large internet audience. And, as I have learned from interviews with adolescents, many other types of online demos aim to distract and seduce viewers. The modes of deployment of contemporary public demonstrations seem in fact to be in sync with the expanding use of the internet, and in particular with the "consumption" of videos through this medium.

ENTERTAINMENT ON THE WEB

Demonstrations of skill in video games seem to have broad appeal on the web. Videos of games as they are being played, often accompanied by multiple comments, have become highly popular.

The video clip *Battlefield 3: Most Epic Killstreak!*, which had been viewed more than 77,000 times on YouTube as of 2017, is a good example.[5] The clip presents an excerpt from the war game *Battlefield* in which a gamer managed to exterminate an impressive number of enemies in a very short time without getting killed himself. Associated comments such as the following indicate the viewers' admiration: "Loved the commentary man. . . . But i think u shouldn't fast-forward some parts. . . . U should talk through all those too. . . . Btw, great video)."[6]

This sequence illustrates the links that sometimes connect demonstrations, competitions, and exploits of various kinds. From this standpoint, it can be compared, for example, to rhetorical contests in which candidates seek to demonstrate their eloquence, their virtuosity in handling language.

Beyond such demonstrations of skill, the internet features all sorts of tutorials that seek to explain in an entertaining fashion how to use video games; these tutorials appear to be substitutes for less engaging and less "intuitive" forms of instruction, such as written users' manuals or prose commentaries.[7] Some are designed for beginners and offer overall presentations of games. This is the case with an introductory guide in French to the virtual construction gaming software *Minecraft*. The opening summary of *Minecraft—Le Guide pour bien débuter—Tutoriel Fr: Episode 1*, which had been viewed more than 1.7 million times on YouTube in 2017, presents the content as follows: "You want to know more about Minecraft? By sharing my adventure with you, I'll explain all the basics for surviving and thriving in Minecraft: artisanship, exploration, construction."[8]

The author of the demo explains in a highly pedagogical tone, with the help of illustrations, how to use Minecraft and how a game develops. He offers advice on choosing options and demonstrates, among other things, how to make an oven, a flashlight, and a pickaxe. The web page displaying this video includes many enthusiastic reactions, such as: "No other youtuber has explained the principle of Minecraft as well as you! Every Minecrafter ought to watch these guides, it's the Minecraft bible!" and "no insults its the only video where im sure i can see it with my parents watching lol."

Other gaming demos aim chiefly to present clever moves; these are addressed to more experienced players. For example, the video *Comment avoir facilement tous les Pokémon rares de Soleil et Lune* (Easy ways to catch all the rare Pokemon in *Pokemon Sun and Moon*), viewed more than 698,000 times on YouTube as of 2017, presents a series of strategies for getting around some difficulties built into the popular game.[9] Its summary adopts a humorous tone to explain how to proceed: "You're lost. You can't find your favorite Pokemon and you can't finish the Pokedex? Well, we're here to save you! In this video I'm going to show you the rarest Pokemon, the hardest to capture from the Pokemon Sun and Moon games, and I'll show you where to find them and how to capture them!"

Another type of demo consists in exhibiting software flaws that allow a player to earn points or finish games without following the rules. In other words, they show players how they can "cheat." Others show where to find "hidden stuff." Video game designers often slip quirky elements into their programs, and passionate gamers seek them out. The exercise is something like hunting for Easter eggs—in fact, that is what the hidden elements have come to be called.

The video *Battlefield 4: Easter Eggs*, viewed more than 236,000 times on YouTube as of 2017, offers an example.[10] This sequence shows a giant shark turning up when ten players approach a buoy lost in the middle of the ocean, and a yeti screaming when a player shoots at a white light located in the heart of a mountain. The reactions posted on YouTube (for example, "I really love your imitation of chewbacca ^^") bring to light the amusement produced both by the revelation of these "secrets" and by the demonstrator's commentary.

Among the various types of demos related to the operation of video games, some are designed to allow web surfers to discover new games. These evaluations of new products on the market or of products about to be released are addressed mainly to potential buyers. The commercial stakes of such demonstrations are high for game producers, for they attract a broad public, sometimes at record levels. The YouTube channel of a famous video game evaluator, PewDiePie, was the most followed in the world in 2017, with more than 16 billion viewings and more than 59 million subscribers. The revenues of the YouTuber rose to $15 million in 2016.[11]

Some videos have a promotional aim that is expressed more or less explicitly by their authors. Demonstrators may be remunerated by the producers of video games, especially in connection with product launchings. They may also be remunerated by YouTube when the number of viewings of their "test videos" goes beyond a certain threshold, according to the number of "views." Thus, for example, a test video of *L.A. Noire* was produced a week after the game was launched by a YouTuber named Benoît-Diablox9. Famous among French gamers for his evaluations, in this sequence the demonstrator showed how the game worked, presented its scenario, and introduced its characters.[12]

Similarly, an evaluation of *Pac-Man: The New Adventures* was produced by the no less famous French YouTuber Frédéric Molas, who appeared on the YouTube channel *le joueur du grenier* (the player in the attic). A specialist in tests of "retro" video games that have been coming back into fashion, Molas adopted a humorous tone to reveal the secrets of this new edition of *Pac-Man*.[13] The associated comments that were posted on YouTube, after more than three million viewings, stressed the amusement with which it was received by many surfers, as the following excerpts suggest:

So much crazy laughing on this video !! Seeing Pacman depressed, in thug mode, under LSD or furious is worth all the gold in the world😆.

I adore jdg [abbreviation for *joueur du grenier*] Super video even if for me the best is still the ones[14] about [the video game] Takeshi's Challenge.

Keep it up its rare to see youtubers with a such a quality of video with an excellent setup! the whole thing without clickbait!

But just imagine if video games had really disappeared??? There wouldn't be jdg or squeezie or any youtuber gaming and shit it would be f*ing boring!!

Done in a rush? Seriously, it's just too good! especially the parts with David Goodenough.

The success of these video game demos is such that they have become objects of parody. The French video *Bref: j'ai joué à Call of Duty* [In short, I played *Call of Duty*], viewed more than 800,000 times as of 2017, proposed a caricature of the game being played, with commentary. The comments posted on YouTube show that a large number of surfers appreciated the humor and had fun watching it, as indicated by reactions such as these: "Huge, after more than a year I'm still not tired of this video it's just sublime :)." And "I adore it! I had a great time ^^ On my list of favorites!"

The realm of video games is of course not the only one in which demos are put to use on the internet. Comparable phenomena can be observed in the cosmetics sector in particular. In France, certain "Beauty YouTubers" have become famous influencers to varying degrees, especially among adolescent girls, by testing cosmetics and conducting tutorials showing how to apply makeup and how to use various products and accessories. These tutorials are entertaining teaching tools. Every gesture receives commentary as it is performed. The steps in a "successful" makeup session are spelled out in detail, and all sorts of tricks are disclosed. The narration is often punctuated with anecdotes and with lavish declarations of affection, usually in a lighthearted tone. Moreover, the associated YouTube comments enable surfers to solicit advice from the demonstrator or from other demonstratees.

As an illustration, let us consider the demo *Tutoriel Maquillage no 30: Comme au bon vieux temps!* (Makeup Tutorial no. 30: Like in the good old days!) produced by Marie Lopez, alias EnjoyPhoenix, a French YouTube star in the realm of beauty products. In this demo, viewed more than 1.8 million times as of 2017,[15] EnjoyPhoenix sums up its content: "Today it's makeup! I've shown you the makeup I'm wearing all the time these days, and I couldn't wait to film this video for you, kind of a 'back to basics'! If you liked it and want more videos like this one, don't hesitate to say so in the comments section, or just go ahead and 'like' this video! I love you loads!"

In this tutorial the YouTuber reveals the makeup she chose at a time when she claims she was suffering from a bout of acne. She explains that it was during the summer and she didn't want to use "too much stuff." The comments that accompany her gestures specify just how she applies the makeup, what products and accessories she has chosen, and how she uses various palettes and brushes. The conversational tone she adopts and her many digressions are part of an ambiance meant to be convivial. EnjoyPhoenix has taken care to film this video in her own bedroom, moreover, with natural light and no voiceovers. She explains that she wanted to make it more "authentic," going back to what she used to do before she became famous and started making more "professional" recordings (hence the title: "like the good old days").

Numerous emotionally charged internet comments indicate appreciation of the entertaining character of this tutorial. This is clear from the examples chosen from among a total of 1,113, to which were added some 42,000 signs of appreciation in the form of "likes" as opposed to 1,000 "dislikes."

I really liked seeing you like that, like in the good old days as you say! It's fun to hear you talking for 18 minutes, it's crazy I never get tired of it ♡ By the way the decor you had at the beginning was really great, it's different, and I really loved it! Actually it'd be cool if you changed your decor because it's like really refreshing I wanna say 😊 anyway I'm just trying to say that I like the change and it's more fun to watch how it's changing. Big kisses Marie I love you ♡💍

hi Marie you could make a video about Elsa's brushes Pls♡ You're brilliant I adore you ♡🔒 Kiss😊

I really love when you change places !!! and then with no voiceovers its so much better cause you explain things directly! I really love you like that!

I love this kind of video, well, I do like the video but the way it's filmed without too much setup I imagine, I have the impression that you're closer to us like that, I really love it. The brightness doesn't bother me, well, you can do what you want but I really prefer this way :)

I prefer the videos like that, without voiceovers, it feels like we're "closer," and it's cool.😊

ENTERTAINMENT, CULTURAL GOODS, FORMS OF ART 209

While the tone adopted in these comments has to be grasped as a function of the modes of interaction proper to the participants, it at least makes clear the two-fold dimension of the public demonstration in play. The demonstratees appreciate both the video's usefulness and the form of entertainment it provides. The tutorial also offers an opportunity to ask for advice and for new demos, as the following reactions indicate:

Hi there Marie, I wanted to ask you if the coloring damages your hair a lot, I think you know quite a bit about this because you went from brunette to blonde! I had some highlight coloring done a few days ago at the hairdresser's and OMG it's CATASTROPHIC !!! The highlights turned yellow with orange streaks I give up I don't even dare go out with it lol! So I was thinking about doing another coloring by myself to hide the disaster (that'll be my second coloring). Do you think I should wait before doing it? I can hardly wait for your answer:*

you could do a video "how to do a good job putting on your mascara" because it's no good I've tried to teach myself to do it but my eyes aren't the same. There's one where the mascara is on rite and the other isn't (help meee!!! pls)

super video as usual, I'll really love it if you make a special one about how to take care of acne and acne scars it'd really be very useful. Kisses

hey Marie could you do a video about makeup with the naked smoky? 😏 If not super video and big kisses 💋💗

These reactions offer further evidence that public demonstrations must not be perceived a priori as isolated moves if one is trying to get an overall sense of what they consist of and what is at stake. The celebrity of the YouTube star, the audience she manages to pull in and hold onto, and part of the stakes in each of these videos all have to do with the fact that they are produced in a series.

Some of the stakes are economic. The tutorial we have just examined seems indeed to be an important promotional vector for the various products used. Precise references to each item are indicated on the web page featuring the video. In addition, hyperlinks make it possible to access vendors' sites in one click and order the products immediately online.

The commercial basis of the YouTuber's activity is also in play in this serial production of demos. It is important to note that, like the number of viewings on YouTube, the various signs of celebrity and appreciation as well as the traffic directed at online commercial sites can be monetized by means of various mechanisms with the assorted economic actors involved. Thus we can understand why the YouTube demonstrator invited the demonstratees, on her tutorial's web page, to indicate whether they had appreciated the video by leaving a comment or a "like," while also inviting them to subscribe to her YouTube channel (which had more than 2.8 million subscribers at that point) in order to be kept informed about her future presentations.[16] We can also see why links to other sites (i.e., Snapchat, Facebook, Twitter, and Instagram) were included.

I should note that there is a whole range of "beauty" demos on the internet, including video tests of cosmetics in which the pedagogical aspect, teaching viewers how to apply makeup, is less pronounced. Even in these cases, though, there is a demonstrative dimension, to the extent that words and gestures are combined to show how to use the products tested and how the makeup looks on the demonstrator's face. In videos targeting French speakers, the demonstrators' use of anecdotes and gestures of complicity in an often playful tone also aim to entertain the viewers and make the exercise enjoyable to watch. At the same time, these video tests play a more or less significant and explicit promotional role for the products used. The demonstrators are often remunerated in various ways by the cosmetics manufacturers. For the latter, the commercial stakes appear extremely important: as with the videos evaluating video games, the audience is potentially vast.

The French product-testing video *top or flop: test products and give your opinion // good=top // not good=flop,* presented by a young woman called Horia, is a good illustration of these phenomena.[17] In this sequence, the demonstrator begins by explaining how she got a little scab on her chin and tried to hide it with the help of a concealer. Then she shows the effects of various cosmetics applied to her face, one after another, commenting on them in a lively manner. The opening teaser for the video announces what the YouTuber intends to do, and it sets the tone: "Heyyyyy! I'm happy to see you again for a video with no beating around the bush (like always, right 😁). Today I'm going to test a bunch of products that I've never tested before and I'll tell you whether or not they're worth the money! So, top or flop? 🙊"

Numerous comments stress the enjoyable dimension of this sequence, while expressing strong feelings. One viewer writes: "Horia, the one who puts you in a good mood with just a video ♡♡😚😚." However, several elements are present to remind us that we are seeing a video made for commercial purposes: for example, references for the products tested are posted, and links to online vendors make it easy to purchase them. We can also point to the requests for comments and for "likes" that Horia formulates on her page.

This commercial dimension is even more visible in certain "video tests" that are devoted to promoting the products of a single brand. The French video *Huda beauty: How to use the products* is a good example.[18] Its author shows how to put on false eyelashes made by the Huda Beauty company. The demonstrator also shows rather playfully how a liquid lipstick can be used as an eyeliner. This demo produces comments containing expressions of delight and overflowing with affection, as in the following: "I adore it !!! ♡ ♡ ♡ ♡ ♡ ♡ as usual !!!! ♡♡♡♡," or "Hey, Beautiful Sophie, top video as always. . . . I love your naturalness and simplicity!! I heart this brand and the false lashes are really super, the ones in the video are some of my faves, I adore them!! For the liner, it's a super idea, I'll try it right away! Lots of hugs and kisses my lovely."

The examples of YouTube videos we have just examined illustrate how various types of public demonstrations, especially in the realms of video games and cosmetics, are at the heart of a vast audiovisual production on the internet. The size of the audiences concerned also appears considerable. If we compare these demos conveyed via YouTube to the demonstrations we examined earlier, we note that many entertaining demonstrations serve commercial purposes. While some of them are promotional instruments, others constitute spectacles designed to make money on the spot (for example, the Human Beatbox and the hot-air balloon demos). We can thus grasp some public demonstrations as cultural products and as commercial goods intended to be consumed on a more or less broad scale. This point warrants further development.

CULTURAL PRODUCTS AND COMMERCIAL GOODS

Many public demonstrations become objects of commercial exploitation. In the case of street protests, for example, photographers and videographers who capture images of the gatherings may be media employees or may work independently; in either case, they are remunerated for the images they supply to newspapers, television channels, and the like (Barry 2001, 189). These images may then be included in products (here, in the form of reports) sold to consumers (readers or viewers), or they may be used to attract advertising revenue. During this process, a demonstrative production is thus transformed into a spectacle and a media event, images from which are used to lucrative ends.

Various forms of commercialized public demonstrations can be found throughout history. In the early Renaissance, for example, anatomy lessons based on corpses of criminals were spectacles that attracted large, fascinated crowds; the price of admission was often substantial (Van Dijck 2001).

The demonstrations carried out by François Bienvenu in Europe in the late eighteenth and early nineteenth centuries also exemplify this phenomenon of commercialization.[19] Bienvenu began by manufacturing and selling scientific instruments; he started to demonstrate his products in France in order to attract and maintain a clientele. Until the start of the French Revolution in 1789, he showed how to use his instruments in the context of classes, some with entrance fees, but some free of charge. These demonstrations, initially conceived as promotional tools and sources of supplementary income, became Bienvenu's sole source of revenue after the Revolution and until his death in 1831.

In the post-Revolutionary period, then, Bienvenu transformed himself into a peddler of science. He toured widely in the French provinces and abroad, charging fees for attendance at his performances, during which he spread the theories of Lavoisier in particular. His demonstrations were often produced on theater stages in large cities; they were publicized and reported on in the press. Over the years, this itinerant demonstrator adapted his shows to what he perceived as an evolution both in his audiences' expectations and in the field of science. His presentation of theories that were little known outside of Paris, at least initially, came to be intentionally spectacular demonstrations. The more Lavoisier's doctrine was adopted, the more Bienvenu's demonstrations combined entertainment with instruction.

Bienvenu's trajectory thus shows how public demonstrations could serve to spread scientific theories even as they constituted consumer goods, whether in the form of paid lessons or spectacles with entrance fees. But performances of this type could also be inscribed wholly in the context of the development of an entertainment industry; as an illustration, let us look again at P. T. Barnum's exhibits of "mermaids."

As we have seen, Barnum rented a "mermaid mummy" to put on display beside other unusual objects in a museum that he had created in New York.[20] The impresario remained vague on the status of the mummy. He invited museum visitors to decide for themselves whether it was a fake or, on the contrary, an authentic cadaver and thus a demonstration of the existence of mermaids. Barnum sought to attract curious viewers by skillfully planting doubt and by indicating confidence in the capacity of each individual to distinguish fakery from reality, without relying on the opinions of experts. This approach was actually characteristic of a more general practice.

In a century marked by an increasing professionalization of science and thus an increasing exclusion of amateurs from the production of knowledge, various forms of staging and spectacles that engaged and highlighted the audience's faculties of observation grew in popularity. Taking advantage of this evolution, Barnum contributed to the development of an entertainment industry based on two elements: on the one hand, the stimulation of curiosity on the part of the general public, and, on the other hand, the opportunity provided to individuals from different socioeconomic backgrounds to see and wonder at astonishing phenomena in respectable contexts, even though that practice was socially frowned upon (Vidal 2006, 7). The mermaid mummy was thus an exhibit—which spectators could view as a demonstration or not—representing a consumer good in a cultural industry in the context of a market based on curiosity and wonder.

Barnum managed to ensure the commercial success of his products thanks to the ingeniousness he displayed in his stage settings, in the creation of new spaces for his offerings (e.g., a private museum in New York, a circus), and in his press campaigns (Vidal 2006, 79). His recourse to mass communication was an essential element in the development of the entertainment industry.

A contemporary illustration of large-scale commercial exploitation of public demonstrations can be found in an exhibition held at the Château de Versailles in France between October 2010 and April 2011. In this exhibition, titled *Sciences and Curiosities at the Court of Versailles*, one section was specifically devoted to the scientific demonstrations that had taken place in the palace in the past.

Demonstrations presented for the king's benefit were frequent at Versailles; fewer were performed in front of the entire court. Such demonstrations earned immense symbolic credit for their authors, along with opportunities to find buyers for the devices used. These spectacles were conceived as entertainments designed to satisfy curiosity while reinforcing the king's prestige.[21]

The *Sciences and Curiosities* exhibition offered several examples. One involved a demonstration of electricity that had been staged by Abbé Jean-Antoine Nollet in 1746 in the palace's Hall of Mirrors. Nollet, a professor and demonstrator who was a prominent presence at court, invited several dozen people to hold hands; he then transmitted an electric charge through the chain thanks to his "electric machine" (Blondel 2010).[22] Nollet's demonstration aroused a great deal of fascination at the time.

The same exhibition also featured a demonstration that had been held in another gallery in the palace, the Petite Galerie, using a "burning mirror" (Lehman 2010).[23] This performance aroused much curiosity and made an impression on Louis XIV and his court. Thanks to its previously unequalled size (nearly one meter across), simply by reflecting the sun's rays this mirror could generate an intense heat capable of melting all sort of metals, a feature that had already been the object of various public demonstrations.

The demonstration carried out in the Petite Galerie brought to light another of the mirror's properties: how the light generated by a single candle placed in front of the mirror could illuminate the entire gallery. The king showed that he could read a letter from one end of the gallery to the other, thanks to this sole source of light. The demonstration made especially tangible the power of the emblematic star of the Sun King. In 1669, Louis XIV acquired one of these mirrors, made by François Villette, an engineer from Lyon and a fireworks expert at the court, for the price of seven thousand pounds.

These entertaining spectacles were thus sources of various more or less direct remunerations for their authors. The *Sciences and Curiosities* exhibition exploited them anew by presenting the demonstrative apparatuses that had been used originally. This exploitation was commercial in nature: viewers had to pay entrance fees. In addition, related products were offered for sale, especially in print form.[24]

But the contemporary exploitation was also of symbolic value. *Sciences and Curiosities* served to enhance the prominence of various institutions and their representatives, along with its partners and benefactors (e.g., the Museum of Versailles, Samsung, Saint-Gobain, and the Macdonald Stewart Foundation).[25] The exhibition promoted science and history in general to a broad audience while serving at the same time as an occasion for scientific meetings.[26] In addition, it once again offered the possibility of both stimulating and satisfying public curiosity, diverting and edifying audiences. These operations could be carried out on site at the Château de Versailles, but also by way of publications and audiovisual or internet presentations.[27]

This case not only illustrates the way public demonstrations can take the form of cultural products and market goods, it also shows how such demonstrations can be exploited more than once and over extended periods of time. Moreover, it is noteworthy that, as cultural products, public demonstrations can take vastly different forms: as objects presented in museums, street spectacles, or audiovisual productions, for example, and even as works of art. Let us now look at this latter category more closely.

WORKS OF ART

In his famous painting *The Anatomy Lesson of Dr. Nicolaes Tulp*, Rembrandt depicted a dissection being carried out by Professor Tulp before a group of surgeons (see figure 7.3). The public demonstration took place on January 16, 1632, using the body of a condemned criminal who had been hung for theft that very day. The painting has been imitated or parodied by contemporary artists in comic strips (Goscinny and Uderzo [1972] 2004) and photographs[28] as well as in new paintings.[29]

Other artists have appropriated public demonstrations as their subject matter but without proceeding to distort them. Some have made them modes of expression. Public demonstrations have been used by contemporary artists for various purposes, whether as tools or as works of art. Such demonstrations may aim to express critical viewpoints about

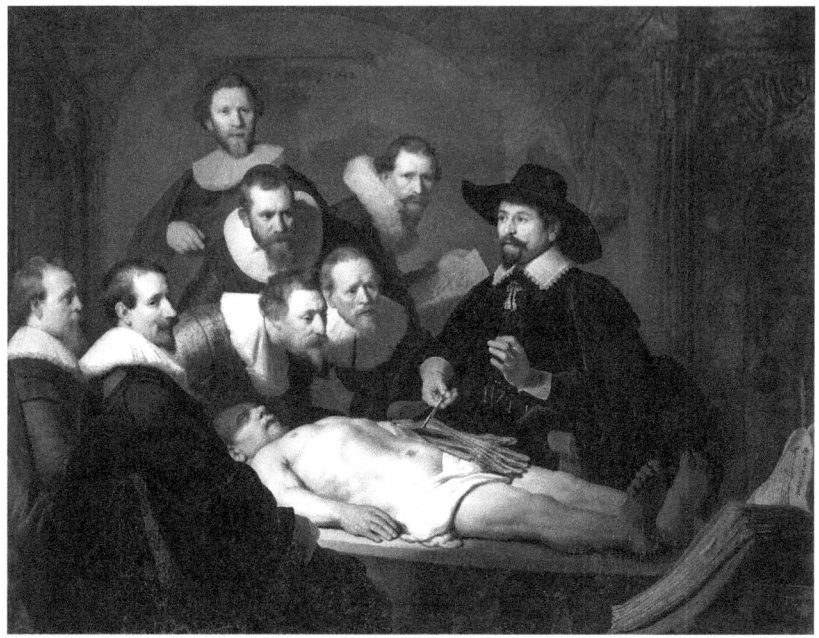

7.3 Rembrandt, The Anatomy Lesson of Dr. Nicolaes Tulp, ca. 1632, Mauritshuis Museum, The Hague.

marketing methods in the domain of new technologies, and by extension to express criticism of capitalism more generally. They may also be used to criticize certain practices in business enterprises or in the art world. These approaches are sometimes developed in essays and theoretical reflections.

Certain contemporary artists also use public demonstrations to deliver various perspectives on the world, sometimes ironic or poetical, and to draw attention to specific phenomena.[30] Their performances may bear on the place of the body in technological culture, for example, on the nature of digital objects, or even on the uses of demos. These artists may seek to entertain, to spur reflection, or to inform the public. These points call for further elucidation.

As an illustration of the way public demonstrations can be mobilized in the context of a critical project, let us look at a series of works by David Byrne focused on PowerPoint demonstrations, presented in *E.E.E.I.* (*Envisioning Emotional Epistemological Information*), a book that came with a DVD (Byrne 2003).[31] The artist critiques in particular the glorification of collaborative efforts in the world of business and beyond, efforts crystallized in meetings where PowerPoint is used. According to Byrne, the imperative to collaborate results in uniformization and conformity.

Some of Byrne's own works consist in PowerPoint demonstrations, and in images created with the help of that software. These works mock the aesthetic decisions that are generally made for such presentations, in particular the color choices and the explanatory schemas used. Slides characterized by overly complex graphics, information overload, and saturated color palettes shine an ironic light on the composition of the PowerPoint demonstrations commonly produced for business purposes.[32]

In his *Discontrol Party 2*, artist and researcher Samuel Bianchini formulates a critique of the increasing control exercised over citizens with the help of new technologies. He seeks to demonstrate that such control can be avoided if individuals unite to prevent it. Bianchini's public demonstration took place in 2011 in Paris at the Gaîté Lyrique, a center for digital arts and modern music.[33] A large number of people had been invited to dance in this former theater, transformed for the occasion into a surveillance center. In order to keep track of the attendees' movements, all sorts of technologies were brought into play (including cameras, radio frequency identification devices, and facial recognition tools). The data relating to the guests' positions, collected by the various sensors, along with the analyses made by the central surveillance system, were displayed on large screens set up around the edge of the dance floor. The dancers and others could thus follow their own movements and observe the impact of their behavior.

As the evening progressed, the devices proved to be increasingly less effective owing to coordinated actions by the guests. As more individuals became involved and as they came together to dance and to embrace, following them became more difficult: the digital capabilities turned out to be inadequate, and perspiration made some bodily sensors inoperative. The event thus demonstrated that the behavior of social actors could combat the most sophisticated technologies of control.

In other critical approaches, certain projects have condemned the contemporary capitalist system by denouncing the way in which individuals are compelled to present their work frequently and rapidly, commonly in the form of demos. According to Lorne Falk and Heidi Gilpin

(1995, 134), this imperative, which is felt most directly by practitioners of the new technologies, has contaminated the art world. Just as frequent brief demos of projects in preparation are generally required in the context of new product development, artists who use new technologies are also under increasing pressure to produce demonstrations of unfinished electronic works. Demos of prototypes are increasingly accepted and highlighted in art galleries, to the detriment of the sort of immersive experience that can result from displays of completed work. The aesthetic of the demo that has developed in this direction ultimately disrupts the creative process (134).

A similar critique has been developed by authors who have denounced the tyranny of "demo or die" in the art world. According to Peter Lunenfeld, for example, the promotional and evangelistic aims that characterize demos in the information technology industry have been adopted by artists and designers who use the new technologies in their work (Lunenfeld 2000). These creators are increasingly compelled to produce demos along similar lines if they are developing interactive projects, and especially if they do not produce digital images alone. Such demos can be circulated quite easily and can be exhibited without the artists' presence.

According to Lunenfeld, the use of demos is also favored by the evolving pedagogy in art schools. Such schools train artists to produce discourse about their work, to present their works orally, and to interact with the public. These practices tend to create predispositions that may help explain the increasing recourse to demos and to related forms of artistic expression such as lecture-performances. As its name suggests, this latter exercise consists in carrying out a performance and commenting on it at the same time, or else in giving a lecture that is accompanied by a performance.

As an example, let us look at a lecture-performance carried out at the Pompidou Center in Paris by Guillaume Desanges with the help of Alexandra Delage. This event, called "Signs and Wonders," sought to retrace the history of minimal and conceptual art.[34] As Desanges spoke, Delage illustrated his discourse by manipulating geometric forms. These forms were projected as shadow figures on a screen. By arranging them in various ways, Delage managed to bring to light different dimensions of the

ENTERTAINMENT, CULTURAL GOODS, FORMS OF ART   219

compositions of artists who had marked the history of minimal and conceptual art.

As many demos do, this lecture-performance combined words and gestures. It was characterized by the manipulation of devices, not all of which were high-tech. The discourse and the manipulations were mutually reinforcing. To be sure, the event was not a presentation designed to show how a technology functions, except possibly in the most general sense of those terms. But the exercise did convey certain demonstrative elements, as an excerpt from the teaser accompanying the video recording indicates: "Speculation based on a play of coincidences, the lecture is entirely illustrated by shadow figures, created on stage without virtuosity but with the demonstrative intent to shape the essential contours of forms, light, and darkness."

Along the same lines as this lecture-performance, a certain number of contemporary artists use public demonstrations to propose different types of representations of the world, sometimes but not always in a critical vein. Novmichi Tosa has turned an ironic gaze on Japanese industry through his "product demonstrations."[35] With his older brother Masamichi Tosa, he created an "artistic production unit" called Maywa Denki, borrowing the name of a company run by their father. That company produced electronic tubes for Toshiba and Matsushita Electric until the late 1970s.[36] As "president" of Maywa Denki today, Novmichi Tosa puts on his performances dressed in a uniform typical of the ones worn by Japanese workers in the electronics industry when the country's economy was growing fast.

Tosa calls his works "products," and he calls his presentations "product demonstrations." He specializes in creating the most wildly fantastical musical instruments, showing how they work by adding gestures to oral explanations, and by performing pieces that can be very brief. With dry humor he turns an amused gaze on the modes of promotion of industrial objects. On February 18, 2010, for example, in the context of an exhibition titled *Gadget OK* organized by UCLA in Los Angeles, he performed a demonstration making fun of the large-scale exploitations of product demos in ostensibly "live" form, which consist in recording them in order to post them on the internet. Tosa encouraged the spectators present to film his performance and post it on YouTube. One resulting video has been viewed more than 15,000 times as of 2017.[37]

A certain number of works by contemporary artists refer to the notion of demo in a somewhat offbeat sense, as in "demoscene." This term refers to an international community and an underground culture that brings together small groups of computer specialists and artists to produce audiovisual works, typically based on small software programs. The execution of this type of program constitutes a "demo." In this instance, the term does not refer to a demonstration of how a software program works, but rather to the possibilities opened up by information technology in general, and more particularly to the talents of the programmers, musicians, and graphic artists who have developed the corresponding audiovisual work. These productions are presented and entered into competition during "demoparties" that are organized regularly throughout the world and that sometimes draw thousands of participants.[38] The events that have been organized annually since 2016 by Nick Montfort, alias "nom de nom," under the title "Synchrony," are good examples.[39]

Other uses of demos in contemporary art are illustrated by the work of Étienne Cliquet and the group of artists known as Téléférique, of which Cliquet was a cofounder in 1999 (Cliquet 2010). Twice a year, from 1999 to 2005, the collective met to make demos, in a process that developed gradually over time. The demos were produced in a series of different spaces: an office building, a movie theater, an art school, a university, an abandoned factory, a castle, and a rooftop, specifically the rooftop of Le Corbusier's Cité Radieuse in Marseille.[40] This last venue was chosen for the thirteenth Téléférique gathering, also known as "demo 13," held on January 26, 2002. An apparatus featuring a giant portable computer had been set up to produce demos at night. Illumination from the screen extended the light coming from the city below, as can be seen from photos taken during the event.[41] This gathering was followed by others, organized during a six-day period in Paris, in an artist-run exhibition space called Immanence, in February 2002.[42]

The demos produced in this context were conceived as a new form of meeting, and as a way of establishing links among artists and other individuals, different from the kinds of meetings fostered by exhibitions or conferences. Various people had been invited to create their own demos, instead of attending as passive spectators. A CD-ROM containing software programs made by members of the collective had been distributed to preselected participants during interviews; the CD served as an invitation and as a means for artists to present themselves and their work. These private meetings themselves mobilized demos, conceived as follows: "[We are to] bring our own laptops, Mac or PC, to someone's home or workplace, spark a human connection, set up on a table, play, show, discuss, and leave."[43]

All these demos were conceived as a fluid, informal, non-alienating way to enter into a relationship with others on an egalitarian basis. In this respect they were associated with what was perceived at the time as the new connected world of the internet. They were also viewed as sources of gatherings similar to those associated with concerts, where spectators can take pleasure in anticipating the musicians' arrival. In the case of the Téléférique demos, this anticipation could be related for example to the time needed to restart a machine before the beginning of a performance.

The demos were understood, moreover, as means for deploying software programs envisioned either as tools or as works in their own right. Demos represented a way for these programs to be fully realized and made visible. But this actualization did not come about in a solitary framework, as is the case with many uses of software programs. During these demos, each program was deployed, on the contrary, at a moment when a relationship was being established.[44]

Some of the demos Cliquet presented during these collective sessions consisted in brief animations created with the help of screen-capture software. Such software is frequently used to create demos of programs, especially in the domain of video games. The animations feature icons and graphic symbols that move about the screen, as in a couple of works created in 1999 and posted on Vimeo, "Propriétaire"[45] and "Un tour de métro."[46] These sequences took the form of executable computer files developed in reaction to current events or other more personal situations (Cliquet 2010).

The demos of this artist echo one of his own reflections on the relationships and exchanges that have prevailed for many years, starting before the Open Source movement, among certain groups of software users such as DECUS or SHARE, groups sometimes known as "Birds of a Feather."

Cliquet sees these collectives as functioning in an informal mode marked by horizontal relations; within the groups, demos constitute occasions for meetings, presentations of one's work, and collaboration. In his view, these groups reflect what "demo 13" sought to achieve. Even if the relations that develop within the collectives on the internet are often less egalitarian and informal than they may appear at first glance (Rosental 2000), we can see how Cliquet's work sheds a favorable light on the demos. These demos appear useful for the successful functioning of ideal microsocieties. In other words, they turn out to support the foundation of utopias, in the original sense of the term.

Cliquet's work also echoes a reflection on artistic practices. Deeming that too much importance is attached to discourse and theory in art schools, he appreciates the possibility in demos of combining gestures with speech. He thinks that exhibits are not effective in bringing to light the substance of digital tools, which are constantly being manipulated and commented on, after having been downloaded and configured. Demos, in contrast, make it possible to reveal their true nature by exhibiting processes, as was done in "demo 13."

Moreover, Cliquet considers that demos raise the question of how to determine precisely what can be considered a work: is it the tool that supports the demo, the demo itself, or what the demo is capable of producing? He notes that demos can engender "alibis." For example, drawings created during performances with the help of digital tools are not necessarily preserved at the conclusion of the demo. Nevertheless, these products can be considered works in their own right.

The artist and software designer Adrian Ward has deployed a similar problematics around the Auto-Illustrator software he developed starting in 2002, and around the demos he produced with this software.[47] Inspired by the celebrated Adobe Illustrator software, Ward's program differs from the model by including functionalities designed to be in part absurd and uncontrollable. As a result, it is hard to determine to what extent the graphic creations produced with the help of this software are the work of the user as opposed to that of the program.

Looking at demos of Ward's software, we may well ask, on the one hand, whether it is Auto-Illustrator itself, the performance of Auto-Illustrator, or the graphic realization produced during the performance that constitutes the work, and, on the other hand, to what extent Adrian Ward is really the author of those products. In fact, as a user of the software, the artist is in part the author of the representations produced. But the program, with its erratic behavior, may be considered as the artist's coauthor. Still, this behavior in itself can be viewed as having been determined by the artist, in that he also created the software program in question. Thus one could argue that the paternity of the representations produced during the demo can be fully attributed to Ward. Yet this thesis can be questioned if the program is deemed to have a certain autonomy with respect to its creator.

Other reflections and realizations on the basis of demos emerged at an exhibition titled *Mode: demo* organized by the Haute école d'art et de design in Geneva (Geneva University of Art and Design) and held May 5–7, 2010, in the context of a Lift Conference.[48] The exhibition presented sixteen works by different artists, consisting in devices that could be the object of demonstrations. The demos could be carried out by their authors or produced automatically—in other words, in "demo mode." In this configuration, as we have seen, a device demonstrates its own operation autonomously.[49] Moreover, a number of these installations were interactive.

"Linyl: Records of Light," created in 2009 by Ishac Bertran, Natalia Echevarria, Benoit Espinola, and Shruti Ramiah, was one of the works exhibited on this occasion. The installation consisted in colored discs created on the basis of photographs or old pictures, and an old turntable equipped with an electronic device that allowed it to create different lighting schemes depending on the discs selected. The demonstration of this device in operation aimed to make visible the poetry that emerges from old rituals and slow gestures, and, by contrast, the absence of such poetry in the contemporary period with its acceleration of the rhythms of life. The stages in the musical experience indeed appeared numerous and lengthy—one had to choose a disc, wipe it off, place it on the turntable, set the turntable in motion—compared with those involved in the use of a digital reader. The aesthetic of the device and the manipulations thus shed a nostalgic light on an unrecoverable past.[50]

"Touch Me," an installation created by Juliette Sallin in 2008, was also presented during the exhibition. As we have already seen, this work consisted in a carpet-like table covering studded with mushroom-shaped forms and incorporating wires with shape memory and electronic sensors. As soon as a visitor pressed on the fabric, the sensors sent an electric impulse to the wires and the carpet began to move. The installation was set up to demonstrate on its own, in a direct, tactile interaction with the public, how a textile could be endowed with mediating capabilities and become a playful material.[51]

Through these installations and others, the exhibition sought to bring to life the expressive power of demos and to help give new priority to gestures and personal experience as compared to reasoning. It aimed to show how demos could entertain, allow interactions with the public, and offer new perspectives on technologies; at the same time, it sought to show the multiple viewpoints allowed by demos: poetic or ironic, for example. In so doing, its objective was to show that it was possible to move beyond the fatalism of "demo or die," and to make demos not a constraint but an opportunity for artists.[52]

The Australian artist Stelarc has seized such opportunities many times in works that attempt to redefine the relations between the human body, technology, and art. In "Stomach Sculpture," created in 1993, Stelarc had a small object characterized as a "sculpture" inserted in his stomach. The object, made of various metals, was capable of expanding and retracting, and also of emitting sound and light. These functions could be activated from outside the body and could be filmed by an endoscopic camera.

The artist had gone to a clinic to have the sculpture placed in his stomach. He had made a demo filming several minutes of the object in operation before an audience consisting primarily of his assistants and a photographer.[53] The video was then presented along with the sculpture in 1993 at the National Gallery of Victoria in Melbourne, Australia. Using the video of the demo, Stelarc showed how the human body could constitute a space for art and technology on the same basis as an exhibition hall. In other words, he showed how the body could become a simple container for a work of art and technology, and how, viewed as a space, its status could shift from private to public.

"Ping Body," a performance carried out on April 10, 1996, during the Digital Aesthetics Conference organized in Artspace in Sydney, Australia, offers another illustration of Stelarc's use of demos.[54] On that occasion, the artist demonstrated in public how his body could be moved not by

ENTERTAINMENT, CULTURAL GOODS, FORMS OF ART 225

the commands of his brain alone, but by information flows exchanged on the internet. In other words, the demo showed how the action of various body parts could be controlled by a technology and the actions of its users. To that end, the artist had had electrodes placed on certain of his muscles. Connected to a complex apparatus, the electrodes emitted electrical impulses in response to signals, called "pings," generated by information exchanges taking place on the internet. Stelarc's body then executed an involuntary choreography, becoming properly speaking a "Ping Body."

As in "Stomach Sculpture," the "Ping Body" demo presented the operation of a particular device. But it also demonstrated publicly how the human body could be placed at the service of art and be inscribed in the extension of a technology to form a hybrid being with that technology. In this configuration, however, the technology did not constitute a bodily prosthesis. The body appeared rather as a potential appendage of the technological apparatus, and as a machine controlled by that apparatus.

The performance also referenced one of the meanings of "public demonstration": a public proof of feasibility. In this instance it was a matter of proving in a tangible way that another world could come into being, one in which humans would be no more than instruments of technology. By that very token the demo was playing one of its customary roles, that is, staging a virtual world.

In the final analysis, we now have access to a broader vision of the ways in which public demonstrations are conceived and realized, in all their diversity. In this chapter we have seen, first, that such performances can have the character of diverting spectacles and can, on this basis, play a number of roles. Even as they entertain, they can also serve to attract a clientele, to highlight skills or products, and to generate commercial transactions. In some cases, such spectacles also constitute the basis for playful activities and for modes of socialization, as well as pleasurable forms of learning. Moreover, they can be sources of gatherings and celebrations, and vehicles for individual, institutional, or collective promotion.

To the extent that certain public demonstrations are exploited as spectacles for the purpose of monetary gain, we have seen that they can also serve as market goods and as cultural products. Finally, we have seen in what ways these performances can be conceived as privileged objects, means of expression, or works of art. In this context we have seen how contemporary art offers a variety of perspectives on these practices—sometimes approving, poetic, or utopian, and sometimes ironic, skeptical, or openly critical.

The representations of public demonstrations proposed by contemporary artists thus constitute objects of study in their own right. At the same time, they raise stimulating questions that invite sociological analysis—important questions about the nature of what we can call the demonstration society. Let us pursue this point, in conclusion.

CONCLUSION

PRACTICING PUBLIC DEMONSTRATIONS

In the course of my analysis, I have shown how a multitude of actors—including scientists, street protestors, salespersons, and many others—could proceed to conceive, prepare, and carry out public demonstrations. By analyzing the unfolding of such demonstrations in a number of cases, historical and contemporary, I have brought to light the scenarios, the stagings, and the equipment that these events could entail. I have examined numerous ways in which demonstrations combined speech, gestures, and devices. I have looked at the extent to which given demonstrations were positioned at the intersection between fiction and reality, at the degree to which they involved emotional work, and at the ways their authors attempted to control audience interpretations. I have also studied the ways in which given demonstrators sought to manage interactions with, and participation by, their demonstratees.

It has thus become clear that public demonstrations can take highly diverse forms, and that the ways in which they are prepared and carried out are no less diverse. In this realm, some highly creative approaches come into view. Given the diversity of the practices I have encountered, it would be counterproductive to propose a summary that would obscure their differences; my goal here is, rather, to affirm these differences.

Nevertheless, some common features can be underlined, and I can offer a synthesis of my results.

In particular, a distinction can be drawn between "closed" and "open" public demonstrations. Those in the "closed" category anticipate or allow very little public participation, if any, and they bear on demands or devices (in the case of demos, for example) that are generally not likely to evolve in the near term. Those in the "open" category, in contrast, authorize interventions by their demonstratees, and in some instances the objects concerned can evolve after the performance.

It also became clear that, while certain public demonstrations are highly autonomous or sufficient unto themselves, others are inscribed in more or less expansive ensembles and take on their full meaning when they are considered as parts of a whole. It is necessary, for example, to distinguish between demos of software that can be carried out readily in a single time span and a single physical space from those featuring much larger apparatuses (as for example in the domain of energy production), which may require performances distributed over time and space as well as multiple support structures. We have observed that even very small objects can call upon significant demonstrative arsenals and be the focus of large-scale demonstrative campaigns. Such campaigns may be deployed by groups of demonstrators acting in coordination, and they may take the form of simultaneous or sequential demonstrations. They may also be engaged in demonstrative confrontations that oppose diverse groups in widespread spatiotemporal domains.

While variations in the ways public demonstrations unfold are numerous, these modes of deployment may nevertheless manifest common features. Whether we are looking at demos or street protests, we can ordinarily observe structured operations, a foregrounding of problems to which solutions are proposed, and significant efforts to calibrate the proceedings in time and space. The formulation of interpretations concerning the actions carried out during demonstrations and recourse to stabilized repertoires are also commonly shared features.

We have seen, moreover, that as a general rule the production of demonstrations is a demanding activity in which much is unpredictable. Public demonstrations certainly do not systematically achieve the results anticipated by their creators. They may encounter reactions ranging from full adhesion to mixed or ambivalent reception to the most virulent critiques. Responses to demonstrations may be all the more unexpected to the extent that the audience does not correspond to the ideal public envisaged or desired by the demonstrators.

Many factors beyond the composition of the actual audience may affect the outcome of a demonstration, moreover. These factors are not limited to the performance of a demonstration in the strict sense. We have seen that actions undertaken before or after a performance, whether these actions are demonstrative in nature or not, can play a determining role. Other contingencies, such as organizational or circumstantial factors, the dynamics of competition, or large-scale political or cultural configurations, or a combination of these, may also help determine the outcome.

In order to pin down the way a public demonstration can be grasped a priori by the demonstratees involved, knowledge of the apparatus on which it relies does not suffice, nor does knowledge about its structure. Demonstrative equipment has limited power, and it is all the less a guarantee of "success" in that the notions of "success" and "failure" are problematic in more than one respect. We have seen, for example, how an avowed failure in the case of a demo could be presented positively as a way to learn and to progress. Contradictory judgments of success and failure can also be produced in the aftermath of a demo or a street protest in order to reinforce the event's effects or, on the contrary, to combat them. Thus the borderline between failure and success drawn by consultants, journalists, or elected officials is not determined solely by the performance under consideration; it may depend on the demonstrative struggles in which it is inscribed.

We have also seen that demonstrators often invest considerable effort and numerous techniques to try to reach their goals and limit their risks. The activities associated with the conception of a demonstration, like those associated with its refinement, are as variable as they are diverse. The preparations may entail tests and iterative adaptations of the demonstrative apparatus, meetings for the purpose of coordination, and rehearsals. They may involve the construction of scenarios and repertoires, the development of stage settings, and a clear division of labor. In certain cases, they encompass actions involving the target audience, either before or after a performance. During the performances themselves, we have seen how demonstrators can resort to a wide range of maneuvers in order to handle bugs and avoid crashes.

In the course of the analysis, we have seen how these skills can be acquired, and how demonstrators' talents vary. In many cases, demonstrators appear to be at least partially self-taught. Acquiring demonstrative skills commonly depends on mimicry. However, it also sometimes depends on prescriptions formulated by chief demonstrators, who are typically experienced practitioners. Normative approaches are abundant in this field; they can be conveyed in the transmission of knowledge and skills from experienced practitioners to beginners, as is often the case, for example, with demonstrators at fairs.

In studying how people learn the art of public demonstration, I have been led to explore some of the social and historical conditions of the production of demonstrations as well as the backgrounds of a number of demonstrators. It has become clear that producing demonstrations can be a job in its own right, but it can also be an occasional activity requiring a specific set of professional skills. In the latter case, we have noted that the roles of demonstrator and demonstratee are neither permanent nor fixed, but can be adopted temporarily and in alternation.

PERFORMANCES AND THEIR AUDIENCES

While examining modes of production and presentation of public demonstrations in various sociohistorical spaces, I have also been led to study the extent to which demonstrations can be apprehended as theatrical performances, and to consider the nature of their audiences. Not all public demonstrations consist in live performances (we know that some, for example, take the form of published texts). It is even problematic to attribute a theatrical character to all live performances of public demonstrations, or to describe them generically as spectacles.[1] Depending on the case, such a status can be contested by certain of the demonstrators and demonstratees concerned, and can be contrasted in particular with the status of public proof. Attributing a theatrical character to a demonstration does not have explanatory power unless one can specify what "theatrical" or "spectacular" elements are involved in the performance, and what might be exchanged in that context.

Nevertheless, it seems pertinent to grasp public demonstrations as theatrical performances in some instances, in particular when there is perceptible recourse to stage settings, the constitution of scenarios, and activity situated at the intersection between fiction and reality. For example, this characterization appears fully legitimate for performances organized in the context of certain artistic projects or certain spectacles with cultural ambitions.

The study of modes of deployment of demonstrations also brings to light the plurality of their audiences. These audiences cannot be reduced, generally speaking, to those of demonstrations as defined and described by Erving Goffman. They are not always constituted by isolated and interchangeable individuals. Their members may be united by diverse relationships, and they may interact among themselves during performances. While audiences may appear to be constructed in certain respects by the performances themselves, they can also be structured at other levels. Demonstrators can take the composition of their audience into account and can manage their interactions by playing on the disparate identities of the spectators and the groups they may constitute.

Furthermore, demonstratees are by no means always passive or locked into roles limited to acclamation. They may be active, reactive, and critical participants. Their own competencies may in some cases be equal or even superior to those of the demonstrators. Some may be in a position to evaluate astutely the scope and limitations of the demonstrations they are watching; they may be able to grasp the fictional ploys and the conventions put into play in the context of performances, and they may produce counterdemonstrations in turn. Given these possibilities, it does not appear legitimate to reduce public demonstrations to powerful tools for manipulation that would deprive audience members of their critical faculties.

The fact remains, nevertheless, that audience members are not always in a position to be fully cognizant of the resources at work in demonstrations, and they may often be mystified, at different levels. A dissymmetry between the postures of demonstratee and demonstrator may stem from a variety of causes. Demonstrators may be equipped with "black boxes," having perfected strategies to orient audience attention toward certain elements rather than others, or they may have carefully prepared responses to possible objections. Demonstrators may possess skills and knowledge that audience members lack. In addition, their manipulations may rely on subtle omissions, ambiguities, and more or less opaque statements. At the opposite pole from a crude undertaking based on brute force, a demonstration may well be based on subtle dialectics and suggestions.

Finally, we must note that notions of convention and manipulation do not exhaust the complexity of possible relations between audiences and demonstrations, any more than do the notions of fiction and reality. The need to go beyond the Goffmanian perspective thus appears all the more compelling.

A DEMONSTRATION SOCIETY

By studying multiple examples of public demonstrations, we have been able to observe the pervasiveness of these operators within societies, throughout history as well as in the contemporary period. It would seem that individuals in today's societies, like those in the past, are or have been confronted in their daily lives with a multitude of public demonstrations in diverse forms, whether or not they are, or have been, the authors of such occurrences. The extent of the phenomena suggests that we are all living in a demonstration society today, and that this was also the case in the past.

I am using the term "society" here in the singular, given the omnipresence of public demonstrations in all the sociohistorical spaces where I have pursued my investigations. However, it would seem legitimate to use it in the plural to the extent that the nature and uses of demonstrations are not necessarily homogeneous from one society or one period to another. Vast investigations will be required if we are to document such variations and evolutions. These investigations will unquestionably make it possible to specify the meanings to be attributed to the term "demonstration societies," this time in the plural.

However that may be, we may wonder as of now whether a demonstration society (or societies) is (or are) organized according to the principle "demo or die," to borrow the slogan of the founder of the MIT Media Lab. At the conclusion of this inquiry, the situation does not seem quite so dramatic. If we base our analysis on the assorted cases studied here, demonstrating in public does not generally involve questions of life and death. The fact remains that the stakes of such a practice appear to be multiple and often extremely high for the actors themselves, and there does exist a demonstrative imperative that comes into play on various grounds. This imperative is well summarized in the expression Q.E.D., *quod erat demonstrandum,* Latin for "what *had* to be demonstrated." This expression, often used to conclude a demonstration of a result in mathematics, illustrates the character of necessity that may be attached to demonstrative operations, given their stakes (e.g., bringing truths to light). This imperative allows us to understand in part why individuals devote so much time and energy to producing public demonstrations or attending them.

The cases we have studied show that demonstrations have been and still are abundantly employed to "lead the world," to borrow an expression from Alexis de Tocqueville (Tocqueville 2000, 435). They serve in particular to prove, to convince, and to promote, but also to influence and to resist. They are mobilized to develop and sell products and to build markets, and for other purposes as well. We have seen how they can also serve as modes of entertainment, consumer goods, cultural products, commercial spectacles, or forms of art. Public demonstrations can possess a utilitarian vocation and an entertaining character simultaneously. The nature of the many demonstrations that take place in fairs, on the internet, or in the context of certain protest movements shows that these dimensions are not necessarily opposed to one another. Moreover, we have seen how public demonstrations carried out in the realm of art can be of an ironic, poetic, utopian, or critical tenor. Their authors can conceive of them, among other possibilities, as tools for edifying crowds, incitations to reflection, or stagings of a virtual world.

These results take us far away from a narrow Goffmanian frame of analysis, one that understands demonstrations essentially as technical reiterations for utilitarian purposes. Upon analysis, public demonstrations appear as a major, multiform, and multiplex type of interaction. Their nature, their roles, and their signification prove to be as diverse as those associated with other broad categories (e.g., institution, finance, people). The inquiries on which I based my analyses underline in particular the fact that public demonstrations can be reduced neither to proofs made public or deployed in public, nor to exhibitions of athletic or artistic prowess, nor to manifestations of faith, nor to the specific form known as demos.

A demonstration society appears, in addition, to be marked by the multiplicity of spaces in which these exercises are practiced. We can observe public demonstrations in marketplaces (e.g., fairs, open markets, businesses), on the internet, on television, or in showcases for technological innovations. They are used in urban spaces (e.g., street protests, outdoor spectacles, religious processions) and in rural spaces (e.g., demonstration farms), within companies, or on industrial sites. They are deployed in academic spaces, in entertainment venues, and in art exhibits.

Their authors are equally diverse. They include scientists, itinerant peddlers, and athletes; consultants, artists, and pedagogues; video gamers, banking advisors, and YouTubers. We can identify unsung underlings of demonstration (e.g., supermarket employees, interns for marketing events) and also rock stars (technology evangelists like Steve Jobs, or famous YouTuber influencers), not to mention business representatives and company managers.

How can this proliferation of demonstrations in societies be explained? While there are many causes, we can invoke in particular the fact that demonstrations often have to be adapted and repeated. The history of mathematics has shown that proofs often have to be rewritten before they are republished. They can never be received in a stable and lasting form; they are always subject to modification, at varying rates of frequency (Lakatos 1976; Goldstein 1995). This observation seems to hold true for other forms of demonstration, forms that rely on types of support other than writing. We have seen how often public demonstrations can be subject to adjustments, recombinations, and reuses, incorporated into vast demonstrative sets.

If demonstrations are considered in isolation, they can easily be viewed as anecdotal and even insignificant practices. This impression disappears when we adopt a different perspective and look at them more globally. As we have seen, we can then perceive their commonalities and their differences, especially with respect to the nature of the stagings employed, the way interactions are handled, or the techniques used to control interpretations. We can also grasp more easily the ways in which different forms of demonstration can be used in combination, substitute for one another, or be hybridized.

In the cases studied here, we have observed that public demonstrations are not isolated moves, as a general rule. They can be included in chains of action and in campaigns that mobilize demonstrative arsenals. Adopting an all-encompassing approach to these phenomena makes it conceivable that it will become possible to grasp their variations and evolutions. One can thus analyze, for example, the various positions demos occupy in demonstrative arsenals in different social and historical spaces. One can study the contexts in which street actions constitute a privileged form of demonstration for protest movements. One can also ask to what extent a resurgence of recourse to spectacular forms of demonstration for the management of relations among science, technology, and society has been occurring at the turn of the twenty-first century.

This is why it seemed useful to me to consider these phenomena in their full extension, even though this is not the general practice in contemporary sociology (Lahire 2013, 60–69). Given the vast scope of the field of investigation, I have had to make rigorous choices in carrying out my observations, relying in addition on a variety of studies in the social sciences. This process has led both to precise results and to vanishing lines that call for further investigation.

In other words, what I have sketched out at this stage is an open theory and a vast realm of investigation. These are not confined to a specialized field (e.g., the sociology of social movements, the sociology of science, economic sociology, or interactionist sociology); they belong rather to a sociological undertaking in the broad sense of the term. They correspond to total social facts in Marcel Mauss's sense (Mauss [1925] 1990, 100).

Public demonstrations appear indeed to concern the whole of society and its institutions, today as much as in the past. Taken as a whole, these phenomena are at once economic, political, and aesthetic in nature. They belong to the order of cognition, law, and religion. They concern domestic life and the professional world. They involve all social classes. The cases I have studied show that public demonstrations are utilized by centralized authorities, interest groups, organizations, and multitudes of individuals. They are carried out in countless sites of gathering and exchange (fairs, markets, political meetings, and so on). Their preparation and execution can mobilize numerous resources, give rise to significant tensions, and arouse passions. Although they may pass unnoticed, they appear to be essential factors in social life and the source of some of society's most intense moments. Ultimately, they seem to involve on a very broad scale the alliances, the material and symbolic possessions, and more generally the future, of individuals and collectives.

As I indicated at the outset, I do not pretend to any degree of exhaustiveness here. The results I have presented are relatively modest in relation to the vast scope and importance of the phenomena I have tried to bring to light. These phenomena warrant further analysis. Certain forms of public demonstration in particular have been subjected to very little study: I am thinking especially, for example, of public demonstrations of feelings or competencies, demonstrations performed in courts of law, and demos carried out by autonomous machines without human intervention. These types of public demonstrations all invite further questioning and investigation.

From the analytic standpoint, I have concentrated in this study on the ways public demonstrations are prepared and executed. But broader and more systematic studies of their economic and political dimensions remain to be undertaken. Their evolution over time and their role in anthropological exchanges also warrant careful study. In the realm of science and technology, it seems particularly important to examine the roles public demonstrations have been able to play in the production and transmission of knowledge throughout history.

An expansive field of investigation in social science—a field to which the present work seeks to contribute—is thus wide open. The analysis has only begun.

NOTES

INTRODUCTION

1. In English, "Sciences [i.e., here, Scientists] on the March."

2. The *Foire de Paris*, held over several days every year, features large numbers of exhibitors promoting all sorts of consumer goods.

3. A major theater festival takes place every summer at Avignon in the south of France.

4. The expression "technology evangelist" refers to an individual reputed to be able to *convert* large numbers of people to faith in a technological product, in a stance comparable to the one adopted by Christian televangelists, especially in the United States. Steve Jobs was a celebrated example of the type.

5. The sociological approach adopted here aims to go beyond the limits of synchronic analysis by a fully historical representation of the phenomena studied.

6. See *Dictionnaire de l'Académie Française*, 9th ed., https://www.cnrtl.fr/definition/academie9/demonstration; *Dictionnaire Larousse* 2015, http://www.larousse.fr; and *Oxford Dictionary* (Oxford University Press, 2015). It is worth noting that the German term "Demonstration" has a similar set of meanings; see http://de.thefreedictionary.com/Demonstration.

CHAPTER 1

1. As Goffman (1974, 59) notes, "these run-throughs are an important part of modern life yet have not been much discussed as something in their own right by students of society."

2. "Although the demonstrating of something can be radically different from the doing of that something, there is still some carry-over—especially if 'real' equipment

is used—and this carry-over can be sufficient to prohibit demonstration" (Goffman 1974, 67).

3. To be sure, this specific way of representing demonstrations raises other problems. In particular, it does not allow us to go in the other direction and conceive of them as a major type of interaction with diversified forms and substantial stakes, as we shall see.

4. For an example in pictures, see https://www.youtube.com/watch?v=Xukqv6zPYP8, accessed December 19, 2017.

5. In what follows, I am relying in particular on the work of Joseph Lampel (Lampel 2001).

6. For a classic example, see Tarde [1890] 1903.

7. Certain definitions of demos, such as Andrew Barry's, also stress these aspects: "If the older scientific or mathematical notion of demonstration implies proof, the idea of the demo can also imply provisionality. A demo model is a display of the possibility of a real object, rather than its actualisation. It is a way of showing what can or might be done" (Barry 2001, 177–178).

8. See https://www.youtube.com/watch?v=y_umsd5FP5Y and https://vimeo.com/728 3341 (both accessed on December 19, 2017); and the analysis of this work by Peter Lunenfeld (Lunenfeld 2016, 349–354).

9. For a comparison, see Molinié 2016 on the case of celebrations in Seville.

10. I am using a pseudonym to protect the anonymity of the actors.

11. See also Le Velly 2007 on this distinction.

12. This work was presented in the context of the exhibit *Mode: demo*, organized by the Haute école d'art et de design (HEAD) in Geneva, May 5–7, 2010, at Geneva's Centre International de Conférences.

13. "Any coding girls here at #mstechdays? Let's have a girls' get-together to get acquainted and chat. Meet tomorrow in front of Goutu, main floor, at the lunch break?" "Coding girls up for chatting over a glass of wine at #mstechdays?" "I'm in! Having trouble finding women coders to talk to at #mstechdays." Translated from the French; see http://twitter.com/hashtag/mstechdays?src-hash, accessed February 16, 2015.

14. Witness selection may also be based on gender. According to Fraser 1992, the role of witness was conceived above all for men; the terms "testimony" and "testicle" have a common etymology, both being derived from the Latin *testis*.

15. On the adaptation of performances to their audience, and the notion of "recipient design," see Sacks, Schegloff, and Jefferson 1974, 727.

16. See www.sundhed.dk. My account here is based on the analyses of Winthereik, Johannsen, and Strand 2008.

CHAPTER 2

1. Microsoft's TechDays, CES, CHTF, MWC, and Vivatech are among current or recent examples.

2. I am using a pseudonym here to preserve the anonymity of the actors.

3. A similar analysis could of course be undertaken into the norms that come into play in the organization of street protests.

4. Technologies presuppose, constrain, and make possible certain behaviors on the part of their potential and actual users (Akrich 1992).

5. See https://guykawasaki.com, accessed May 18, 2017. See also Kawasaki and Moreno 2000.

6. See https://guykawasaki.com/how_to_be_a_dem/, accessed May 18, 2017.

7. To emphasize this specificity, Gallo created a comparative table of the linguistic registers used by Steve Jobs and by Microsoft's former CEO Bill Gates in the context of their public talks (Gallo 2010, 115).

8. http://www.mahalo.com/how-to-make-a-demo, accessed February 27, 2012.

9. A recording of this demo was available on https://www.m6boutique.com/minceur-fitness/amincissant-raffermissant/autre-textile-minceur/basketonic.htm, accessed November 16, 2018. I am offering a retranscription here. Another version of this demo is available on https://www.m6boutique.com/promos/focus-produit/basket-minceur-basketonic.htm, accessed January 17, 2021.

10. See especially Vinck 2011. By construction, the use made here of the term "equipment" comes closer to Dominique Vinck's notion of intermediary object, taken in its initial, most general sense, than to the notion Vinck associates with the concept of equipment, which for him refers more specifically to an "element added to beings (intermediary objects most notably) making it possible to link these to conventional supports and to spaces of circulation" (Vinck 2009a, 66).

11. This model is in fact not very different from the Ciceronian conception of rhetoric as a resource for teaching, moving, and pleasing all at once, or from Horace's representation of poetry as a didactic tool, a means of persuasion, and a source of pleasure (La Grandeur 2003).

12. "In-Store Demo-Event Specialist," in Des Plaines, Illinois, http://www.findjobz.com/job/1599307/, accessed July 20, 2014.

13. See http://offre-emploi.monster.fr/, accessed July 20, 2014.

14. "Demonstrator-Vendor," in the Rhône-Alpes region, http://www.anmv.fr/offre-emploi-commerciale.html, accessed July 20, 2014.

15. See http://www.indeed.fr/cmp/techsell/jobs, accessed July 20, 2014.

16. See also the study carried out on this contest presented in Frances 2015.

17. Marie-Charlotte Morin, "Mt-180-Ma thèse en 180 secondes: en route pour la finale nationale!," interview by Claire Deboves, *CNRS hebdo*, June 6, 2014; http://intranet.cnrs.fr/intranet/actus/140606-mt180-marie-charlotte-morin.html, accessed

June 6, 2014; https://www.youtube.com/watch?v=p01rPKWqI6A, accessed January 27, 2017.

CHAPTER 3

1. See https://twitter.com/search?f=realtime&q=mstechdays%20since%3A2015-02-10%20until%3A2015-02-11&src=savs, accessed February 16, 2015.

2. The exchanges have not been reproduced in full in order to preserve the anonymity of the actors and for conciseness.

3. Ethnomethodologists have thoroughly analyzed the way a social order can be produced on the interactionist level, in particular when confronted with disruptive situations of all sorts (Garfinkel 1984).

4. The event was covered by regional and national print media (*Le Monde, Le Figaro, Libération, L'Humanité, Le Nouvel Obs*, and so on) and by various French radio stations and television channels (France 3, BFMTV, France Info, France Inter, France Culture, and others).

5. The extent of police involvement in street protests varies, of course, with the context. In particular, large-scale events like the one we have been examining have to be distinguished from smaller demonstrations, from those with more violent aims, or from those that take place without advance collaboration or in open opposition to the security forces (Hindle 2006).

6. For analyses bearing on other sorts of demonstrations of strength, see especially Mukerji 2009 and Cossart and Taïeb 2011.

7. For a sociohistorical approach to the profession of attorney, see Karpik 1995.

8. From this standpoint too, the procedure adopted in this work differs from the general dramaturgic approach to daily life developed in Goffman 1959.

9. My discussion here is based on Marie Thébaud-Sorger's work (Thébaud-Sorger 2009).

10. Article published February 10, 2015, at http://www.rtl.fr/actu/economie/, accessed February 19, 2015.

11. Here too my observation corresponds to others made about various types of public meetings. For example, some media accounts of presidential visits present the sensory environment preceding these events as exhilarating to the public, whereas first-hand observations show considerable individual variations in spectators' reactions (Mariot 2001).

12. Frédéric Pereira, "TechDays 2015: Des infographies pour les trois jours de l'événement," February 13, 2015, https://www.fredzone.org/, accessed February 19, 2015.

CHAPTER 4

1. Studies on cognition provide many examples: see for example Quéré 2008 on the topic of trust.

2. From this standpoint, the approach was analogous to the one required, for instance, for a study of the actions involved in building and managing large databases of digitized information (Bowker and Millerand 2008).

3. For insights into the history of artificial intelligence, see especially Edwards 1996 and Pratt 1987.

4. See also the case of product launching discussed in Simakova 2010.

5. On the importance of documentary practices in the management and coordination of major technoscientific projects ("big science"), see Chompalov and Shrum 1999.

6. This program was managed by the Directorate General for Telecommunications, Information Market and Exploitation of Research of the European Commission from 1994 to 1998. It was followed by two other European programs: "Information Society Technologies" (1998–2006) and "Information and Communication Technologies" (2007–2013).

7. On the misuses of quantitative indicators in the field of research evaluation, see Gingras 2016. In comparison, on the strengths and limitations of qualitative evaluation of research projects, see Lamont 2009 and Rosental 2010.

8. See the account of the teleconference organized on June 18, 1997, in *ACTS Newsclips*, no. 26, July 1, 1997, http://cordis.europa.eu/infowin/acts/ienm/newsclips/arch1997/970686at.html, accessed December 19, 2017.

9. In the discussion of EDF undertakings, I am relying on Topçu's analyses (Topçu 2013, 103–104, 199–200).

10. For a comparison with the British case, see De Carvalho 2018.

11. Demonstrations of absence during visits to technological installations can be observed in many other cases. To take just one example, we can consider the demonstration of the nonexistence of "N-rays" produced by American physicist Robert Wood on the basis of his visit to the laboratory of French physicist René-Prosper Blondlot in the early twentieth century (Ashmore 1993).

CHAPTER 5

1. On the significant number of accidents attributable to software systems, see MacKenzie 1994.

2. For a sociological analysis of the causes of that catastrophe, see Vaughan 1996.

3. For the study of a comparable dynamic, see MacKenzie 1990.

4. See Tod Machover and Dan Ellsey, "Inventing Instruments that Unlock New Music," TED Talk 2008, https://www.ted.com/talks/tod_machover_dan_ellsey_inventing_instruments_that_unlock_new_music#t-1190664, accessed November 22, 2020.

5. In the developments that follow, I rely especially on Jasanoff's analyses.

6. "Most people in such nations are quite intent on immediate material gratifications, and since they are always unhappy with the position they occupy and always free to abandon it, they think only of ways to change or improve their fortunes.

To minds so disposed, any new method that shortens the road to wealth, any machine that saves labor, any instrument that reduces the costs of production, any discovery that facilitates or increases pleasures seems the most magnificent achievement of the human mind. It is primarily for these reasons that democratic peoples devote themselves to science, understand it, and honor it" (Tocqueville 2000, 436–437).

7. On the interpretive flexibility to which technologies may be subjected, see Woolgar 1991.

8. See Minimal USA, "Introduction Video for 'Making Room' Exhibition," YouTube, February 1, 2013, https://www.youtube.com/watch?v=HUIChQxUkbg, accessed November 22, 2020.

9. Many other examples can be found in demonstrations of technologies intended for the general public; for example, see Collins 1988.

10. Let us recall that, according to Thomas Kuhn (1962), every change of paradigm brings with it a simultaneous revolution in worldview. Thus Uranus, after having been *seen* for nearly a century as a star, then as a comet, was *seen* beginning in 1781 as a planet. For Kuhn, such a transformation in the vision of an object does not stem from an isolated incident, nor is it simply a change in interpretation of a singular phenomenon identified in a stable fashion. This evolution stems from connected modifications in the ways of viewing multiple elements of the world: in other words, a paradigm shift.

11. The demo could be viewed at https://www.youtube.com/watch?v=vvlWcpixo9s, from 24:57 to 32:14, accessed November 8, 2018.

12. See https://gamedevacademy.org/author/david-catuhelive-fr/, accessed November 22, 2020; https://github.com/davrous, accessed January 14, 2021; https://twitter.com/pierlag?lang=en, accessed November 22, 2020.

13. Guillaume Serries, "TechDays 2015, les développeurs Microsoft à l'assaut de la matrice," February 10, 2015; see http://www.zdnet.fr/actualites, accessed February 19, 2015.

14. See https://twitter.com/hashtag/mstechdays?src=hash, accessed February 16, 2015.

15. This phenomenon can be considered in relation to Proustian analyses, in which characters in the narrative often have very different ways of grasping signs that may seem to be bearers of a single interpretation and of a programmed effect (Proust 1927, 567–568).

16. I am relying here on the analyses in Vidal 2006, 69–81.

17. Admittedly, pity worked in the movie's context, but that register was not used in the context of the fair.

18. See https://www.youtube.com/watch?v=KJbCqd3e3v4, accessed August 10, 2017 (341,208 viewings).

19. https://www.youtube.com/watch?v=QOPFg2mo, accessed August 3, 2017 (1,661,665 viewings). See also CNET, "Steve Jobs' Demo Fail," YouTube, June 7, 2010, https://www.youtube.com/watch?v=znxQOPFg2mo, accessed November 22, 2020.

NOTES   243

20. See Guy Kawasaki, "How to Be a Demo God," January 23, 2006, https://guykawasaki.com/how_to_be_a_dem/, accessed May 18, 2017.

21. See https://www.youtube.com/watch?v=lpRdOU9bjo4&list=PLbl2SbVIi-Wq11AoEA2bQ17FZDaylVROE&index=2, beginning at 8:18 and ending at 14:17, accessed February 11, 2018.

22. On the use of notions such as "prototype," "version," and similar terms, see also Collins 1988 and Simakova 2010.

23. On the nature and stakes of debates over nanotechnologies, see also Vinck 2009b and Hubert and Vinck 2017.

CHAPTER 6

1. See http://www.1968demo.org, accessed July 10, 2017.

2. See Jim Grubb, *Cisco Blogs*, https://blogs.cisco.com/author/jimgrubb, accessed July 10, 2017. Grubb's position has been labeled "Chief Technology Evangelist" in Cisco's "Customer Experience Center."

3. On militancy skills in the union context, see Mischi 2013.

4. See http://www.cnrtl.fr/definition/academie9/demonstrateur, accessed November 22, 2020.

5. See Greene 2004, 271–272. The developments that follow are based on Greene's study.

6. For example, Pôle Emploi, a French agency helping unemployed people enter the workforce, and Centre d'Information et de Documentation Jeunesse in France, a French agency providing students with information about working in various trades.

7. See in particular http://www.hotessejob.com/emplois-par-metier, accessed July 20, 2014; and http://www.marketvente.fr/marketing-terrain, accessed July 20, 2014.

8. See http://www.hotessejob.com/emploi/103937, accessed July 20, 2014.

9. See http://jobs-stages.letudiant.fr/stages-etudiants, accessed April 28, 2011.

10. See http://www.indeed.fr/cmp/techsell/jobs/, accessed July 20, 2014.

11. See http://jobs-stages.letudiant.fr/jobs-etudiants, accessed April 19, 2011.

12. http://createdcash.com, accessed July 21, 2014; http://clubdemo.com/careers/, accessed July 17, 2017; and https://www.milwaukeejobs.com/job/detail/24398708/PRODUCT-EVENT-DEMONSTRATOR-PART-TIME, accessed July 16, 2017.

13. See http://www.marketvente.fr/marketing-terrain, accessed July 20, 2014.

14. See http://www.indeed.fr/cmp/Techsell/jobs, accessed July 20, 2014.

15. http://www.anmv.fr/offre-emploi-commerciale.html, accessed May 3, 2011.

16. See http://jobs-stages.letudiant.fr/stages-etudiants, accessed April 28, 2011.

17. See http://www.indeed.fr/cmp/techsell/jobs, accessed July 20, 2014.

18. See http://hotessejob.com/emploi/103397, accessed July 20, 2014.

19. See http://jobs-stages.letudiant.fr/jobs-etudiants, accessed April 19, 2011.

20. See http://jobs-stages.letudiant.fr/stages-etudiants, accessed April 28, 2011; http://www.anmv.fr/offre-emploi-commerciale.html, accessed May 3, 2011; http://www.indeed.fr/cmp/techsell/jobs/, accessed July 20, 2014; http://www.hotessejob.com/emploi/103937, accessed July 20, 2014.

21. http://www.anmv.fr/offre-emploi-commerciale.html, accessed May 3, 2011; http://www.indeed.fr/cmp/techsell/jobs/, accessed July 20, 2014.

22. See http://www.marketvente.fr/marketing-terrain/, accessed July 20, 2014.

23. http://www.huzz.com, accessed December 11, 2010; http://offre-emploi.monster.fr, accessed July 20, 2014.

24. See http://offre-emploi.monster.fr, accessed July 20, 2014.

25. http://huzz.com, accessed December 11, 2010.

26. See http://www.stepstone.fr, accessed December 11, 2010.

27. See http://offre-emploi.monster.fr, accessed July 20, 2014; http://jobs-stages.letudiant.fr/jobs-etudiants, accessed April 19, 2011; http://www.huzz.com, accessed December 11, 2010; http://www.stepstone.fr, accessed December 11, 2010.

28. http://www.lecam-2000.com/pages/inforegltmarches.html, accessed December 4, 2014.

29. On the distinction in French between *démonstrateurs* and *posticheurs*, see also Le Velly 2007, 146–147.

30. See https://www.merriam-webster.com/dictionary/huckster, accessed April 30, 2020.

31. [Translator's note] I have retained the term *posticheur* throughout for want of a sufficiently neutral English equivalent. A posticheur would readily identify himself as such; it seems unlikely that a huckster would do the same.

32. See http://www.cnrtl.fr/definition/aboyeur, accessed November 22, 2020.

33. See http://www.marchesdefrance.fr/histoire-de-la-federation, accessed December 4, 2014.

34. See http://www.larousse.fr/dictionnaires/francais/camelot/12506, accessed November 22, 2020; https://dictionnaire.reverso.net/francais-definition/camelot, accessed November 22, 2020.

35. See http://www.studyrama.com/formations/fiches-metiers/, accessed December 11, 2010.

36. See also Le Velly 2007.

CHAPTER 7

1. The demonstrators are labeled A, B, and C to preserve their anonymity. The interaction is not reproduced in full for conciseness.

2. The discussion that follows is based on the analyses in Thébaud-Sorger 2009.

3. The American Forum Movement consisted in local meetings organized throughout the United States in a format designed to encourage public debate on a variety

of political questions. The discussions took place among members of a "public" that had come together to listen to a speaker.

4. This demonstration took place in August 2014 at Palavas les Flots in southern France.

5. https://www.youtube.com/watch?v=zlq75KsL8, accessed August 10, 2017 (77,806 viewings).

6. The commentaries by web surfers I analyzed seem to proceed less from attempts to manipulate their readers than those observed on commercial sites, where comments can be tied to the production of messages by individuals or companies that may be complicit or competitive with the sales pitch in question.

7. In a reversal of perspective, the capacities related to the notion of "intuition" could be explored in relation to practices such as demos.

8. See https://www.youtube.com/watch?v=o6Kh8XefWfw, accessed August 10, 2017 (1,781,414 viewings). An English-language equivalent can be found at https://www.youtube.com/watch?v=BEH_fMgRNrc, accessed April 4, 2020.

9. See https://www.youtube.com/watch?v=ksRI6jQu5CU, accessed August 10, 2017 (698,084 viewings).

10. See https://www.youtube.com/watch?v=1jt2UoG-9OE, accessed August 10, 2017 (236,660 viewings).

11. See https://fr.wikipedia.org/wiki/PewDiePie, accessed December 31, 2017; https://www.youtube.com/user/PewDiePie/featured, accessed November 22, 2020.

12. See Benoît - Diablox9, "Vidéo-test de L.A. Noire par Diablox9," YouTube, May 26, 2011, https://www.youtube.com/watch?v=CkG0KYMpDvg, accessed January 14, 2021 (394,220 viewings).

13. See Joueur De Grenier, "PACMAN The New Adventures," YouTube, February 11, 2017, https://www.youtube.com/watch?v=l5-FtcHY6QA, accessed January 14, 2021 (7,215,211 viewings).

14. [Translator's note] Here ("the best is still the ones") and elsewhere in these comments, a grammatical error in English reproduces the same type of error in French; elsewhere (for example, in the following comment, "its rare"), a common English error is introduced to hint at the frequency of similar errors in French, not all of which can be reproduced directly.

15. See EnjoyPhoenix, "[Tutoriel Maquillage n°30]: Comme au bon vieux temps!," YouTube, August 21, 2015, https://www.youtube.com/watch?v=SgSAlH6iMTM, accessed January 14, 2021 (2,142,909 viewings).

16. See https://www.youtube.com/user/EnjoyPhoenix, accessed August 10, 2017.

17. See https://m.youtube.com/watch?v=ATxCzm8ibT0, accessed August 10, 2017 (519,645 viewings).

18. See SophieRMakeup, "Huda Beauty // Comment utiliser les produits?," YouTube, November 13, 2016, https://www.youtube.com/watch?v=DlDBmC5rqVE, accessed November 22, 2020.

19. The discussion that follows is based on Bret 2004.

20. In this discussion I am relying once again on Vidal 2006.

21. For a history of the relations between stagings of knowledge and demonstrations of power at the court of Versailles, see Lamy 2017.

22. See http://sciences.chateauversailles.fr/, accessed August 30, 2014.

23. http://sciences.chateauversailles.fr/, accessed August 30, 2014.

24. Besides the exhibition catalog, articles about the exhibit were featured in special issues of the *Cahiers de Science et Vie*: "Versailles: Le pouvoir et la science" and "Les sciences au château de Versailles": see http://sciences.chateauversailles.fr/, accessed August 30, 2014.

25. "le mot de Jean-Jacques Aillagon," http://sciences.chateauversailles.fr/, accessed August 30, 2014.

26. In particular, a colloquium titled "The Court and the Sciences: Birth of Scientific Policy in the European Courts in the Seventeenth and Eighteenth Centuries," held February 3–5, 2011, at the Château de Versailles: https://journals.openedition.org/crcv/11424?lang=en.

27. See especially the "pedagogical resources" on the docsciences website and on the educational website lumni.fr: https://www.reseau-canope.fr/docsciences/Sciences-et-curiosites-a-la-Cour.html; and https://www.lumni.fr/dossier/sciences-et-curiosites-a-la-cour-de-versailles, both accessed November 22, 2020.

28. See for example Anton Solomoukha, *La leçon d'anatomie du docteur Tulp*, 2007, http://www.pascalpolar.be/site/oeuvresview.php?no_inv=solomoukha-o4-020, accessed August 16, 2017.

29. See for example Yiull Damaso's painting, *The Night Watch*, depicting Nelson Mandela's body undergoing dissection: https://www.yiull.com/political?lightbox=dataItem-ihn465v41, accessed November 22, 2020; and https://www.theguardian.com/world/2010/jul/09/nelson-mandela-painting-south-africa, accessed August 16, 2017.

30. For an exploration of the links between the history of aesthetic contemplation and the history of perception, see Crary 2001.

31. See http://davidbyrne.com/explore/e.e.e.i.-powerpoint/about, accessed August 17, 2017.

32. See especially http://davidbyrne.com/explore/e.e.e.i.-powerpoint/photos/still-images/3; and http://davidbyrne.com/explore/e.e.e.i.-powerpoint/photos/tokyo-installation/2, both accessed August 17, 2017.

33. See http://dispotheque.org/fr/discontrol-party, accessed August 17, 2017.

34. See http://www.dailymotion.com/video/xb3n9g_conferences-performances-guillaume_creation, accessed August 17, 2017. See also Cliquet 2010.

35. See http://www.maywadenki.com/biography/, accessed August 17, 2017.

36. See http://www.maywadenki.com/history/, accessed August 17, 2017.

NOTES

37. See https://www.youtube.com/watch?v=VnMkES8O4s4, accessed August 17, 2017 (15,326 viewings).

38. See https://www.demoscene.info/ and https://en.wikipedia.org/wiki/Demoscene, both accessed November 22, 2020.

39. See http://synchrony.nyc/, accessed June 7, 2019.

40. See http://www.teleferique.org/demos/demo13, accessed August 26, 2014.

41. See http://www.teleferique.org/demos/demo13/Marseille, accessed August 26, 2014.

42. See http://www.teleferique.org/demos/demo13/Paris/immanence.htm, accessed August 26, 2014.

43. See http://www.teleferique.org/demos/demo13/Paris/immanence.htm.

44. See the account posted on http://www.teleferique.org/demos/demo13; and http://www.teleferique.org/demos/demo13/Paris/immanence.htm, both accessed August 26, 2014.

45. See https://vimeo.com/18837242, accessed August 19, 2017.

46. See https://vimeo.com/18837454, accessed August 19, 2017.

47. See https://transmediale.de/content/signwave-auto-illustrator, http://artelectronicmedia.com/artwork/auto-illustrator; and https://vimeo.com/17593822, all accessed August 19, 2017.

48. See http://www.ednm.fr/wp-pdf/MODE-DEMO.pdf and http://liftconference.com/lift10, both accessed August 19, 2017. "Since 2006 Lift Events [Leading Innovation For Tomorrow] explores the business and social implications of technological innovation through the organisation of international event series and open innovation programs in Europe and Asia." Cited from http://www.liftglobal.org/content/liftevents, accessed June 25, 2020.

49. See https://www.abstractmachine.net/posts/modedemo, accessed November 22, 2020.

50. See http://ishback.com/linyl/index.html, accessed August 19, 2017.

51. See https://softcircuit.wordpress.com/2010/05/05/preparing-touch_me-for-the-lift-experience-exhibition-mode-demo/; and http://www.ednm.fr/wp-pdf/MODE-DEMO.pdf, both accessed August 19, 2017.

52. See http://www.ednm.fr/wp-pdf/MODE-DEMO.pdf, accessed August 19, 2017.

53. See http://stelarc.org/?catID=20349 accessed August 19, 2017; and stelarcvideo, "Stomach Sculpture," YouTube, April 13, 2010, https://www.youtube.com/watch?v=IFFizqMmlOQ, accessed November 22, 2020.

54. See https://www.youtube.com/watch?v=wTYYJZG0f68; http://www.artelectronicmedia.com/document/stelarc-ping-body; and http://www.artwiki.fr/wakka.php?wiki=StelarC, all accessed August 19, 2017.

CONCLUSION

1. If only on this account, a demonstration society is not the same thing as a society of the spectacle. I should note moreover that in terms of the theses and the methodologies employed, the present work has little in common with that of Guy Debord ([1967] 1994). Let us recall that Debord's political essay, first published in 1967, consisted in a series of aphorisms that constituted a Marxist-inspired critique of capitalism, and especially of commodity fetishism. The notion of spectacle refers essentially to the representations of commodities that purportedly shared in a false vision of the world, structured mediatized social relations, and contributed to the alienation of individuals within the capitalist realm. For Debord, "spectacles," in their diverse manifestations, constitute a propaganda apparatus for capitalism, destined to favor the consumption and reproduction of commodities and to entrench the reproduction of the system and its harmful hold over the lives of individuals; in short, capitalist society essentially constitutes a society of the spectacle. The gulf between the present work and Debord's, then, is easy to see. Not only can public demonstrations not be reduced to spectacles, but also denunciations of manipulations produced with the help of public demonstrations constitute, in this book, an object of investigation in their own right. Finally, my undertaking does not consist in a free-form dissertation but in an analytical work nourished by the results of systematic empirical investigations.

REFERENCES

ACTS. 1997. *Results, Impact and Exploitation.* Interim Report, European Commission, DGXIII/B Ref: AC 1997/1339, May 15.

Akrich, Madeleine. 1992. "The De-scription of Technical Objects." In *Shaping Technology/Building Society: Studies in Socio-technical Change,* edited by Wiebe E. Bijker and John Law, 205–224. Cambridge, MA: MIT Press.

Akrich, Madeleine, Michel Callon, and Bruno Latour. 2002a. "The Key to Success in Innovation Part I: The Art of Interessement." *International Journal of Innovation Management* 6, no. 2: 187–206.

Akrich, Madeleine, Michel Callon, and Bruno Latour. 2002b. "The Key to Success in Innovation Part II: The Art of Choosing Good Spokespersons." *International Journal of Innovation Management* 6, no. 2: 207–225.

Anonymous. 2013. "La démonstration-vente ne s'invente pas." *Le Républicain Lorrain,* October 6. http://www.republicain-lorrain.fr/moselle/2013/10/06/la-demonstration-vente-ne-s-invente-pas.

Aranguren, Martin. 2015. "Emotional Mechanisms of Social (Re)production." *Social Science Information* 54, no. 4: 543–563.

Ashmore, Malcolm. 1993. "The Theatre of the Blind: Starring a Promethean Prankster, a Phoney Phenomenon, a Prism, a Pocket, and a Piece of Wood." *Social Studies of Science* 23: 67–106.

Auray, Nicolas. 1997. "Ironie et solidarité dans un milieu technicisé. Les défis contre les protections dans les collectifs de hackers." In *Cognition et information en société,* edited by Bernard Conein and Laurent Thévenot, 177–201. Paris: Éditions de l'EHESS.

Barry, Andrew. 2001. *Political Machines: Governing a Technological Society*. London: Athlone Press.

Bazerman, Charles. 2002. *The Languages of Edison's Light*. Cambridge, MA: MIT Press.

Bean, Jonathan, and Daniela Rosner. 2013. "Demo or Die? The Role of Video Demonstrations in the Public Domain." *Interactions*. http://goo.gl/XS70ff.

Blondel, Christine. 2010. "L'expérience d'électricité dans la galerie des glaces." In *Sciences et curiosités à la cour de Versailles*, edited by Béatrix Saule and Catherine Arminjo, 237–239. Versailles: Rmn éditions / Établissement public du musée et du domaine de Versailles.

Bloomfield, Brian P., and Theo Vurdubakis. 2002. "The Vision Thing: Constructing Technology and the Future in Management Advice." In *Critical Consulting: New Perspectives on the Management Advice Industry*, edited by Timothy Clark and Robin Fincham, 115–129. Oxford: Blackwell.

Bourdieu, Pierre. (1980) 1990. *The Logic of Practice*. Translated by Richard Nice. Stanford, CA: Stanford University Press.

Bourdieu, Pierre. (1998) 2002. *Masculine Domination*. Translated by Richard Nice. Stanford, CA: Stanford University Press.

Bowker, Geoffrey C., and Florence Millerand. 2008. "Metadata, trajectoires et 'énaction.'" In *La cognition au prisme des sciences sociales*, edited by Bernard Lahire and Claude Rosental, 277–303. Paris: Éditions des Archives Contemporaines.

Brand, Stewart. 1987. *The Media Lab: Inventing the Future at MIT*. New York: Viking.

Braudel, Fernand. (1967–1979) 1992. *Civilization and Capitalism, 15th–18th Century: The Wheels of Commerce*. Translated by Siân Reynolds. Berkeley, CA: University of California Press.

Bret, Patrice. 2004. "Un bateleur de la science. Le 'machiniste-physicien' François Bienvenu et la diffusion de Franklin et Lavoisier." *Annales historiques de la révolution française* 338: 95–127.

Brian, Éric. 2001. "Citoyens américains, encore un effort si vous voulez être républicains!" *Actes de la recherche en sciences sociales* 138: 47–55.

Brook, Peter. 1968. *The Empty Space*. New York: Atheneum Publishers.

Byrne, David. 2003. *E.E.E.I. (Envisioning Emotional Epistemological Information)*. New York: Steidl/Pace/MacGill Gallery.

Cahour, Béatrice, and Barbara Pentimalli. 2005. "Conscience périphérique et travail coopératif dans un café-restaurant." *Activités* 2, no. 1. http://journals.openedition.org/activites/1578.

Callon, Michel. 2003. "Quel espace public pour la démocratie technique?" In *Les sens du public. Publics politiques, publics médiatiques*, edited by Daniel Cefaï and Dominique Pasquier, 197–221. Paris: Presses universitaires de France.

Callon, Michel, and Fabien Muniesa. 2007. "Economic Experiments and the Construction of Markets." In *Do Economists Make Markets? On the Performativity of Economics*, edited by Donald MacKenzie, Fabian Muniesa, and Lucia Siu, 163–189. Princeton, NJ: Princeton University Press.

Capelle, Camille. 2012. "Échanger, concevoir, innover: Analyse ethnographique d'évaluations pédagogiques avec les technologies numériques." Ph.D. thesis. Montpellier: Université de Montpellier 3.

Cefaï, Daniel. 2007. *Pourquoi se mobilise-t-on? Les théories de l'action collective*. Paris: La Découverte.

Chalvon-Demersay, Sabine. 1999. *A Thousand Screenplays: The French Imagination in a Time of Crisis*. Translated by Teresa Lavender Fagan. Chicago, IL: University of Chicago Press.

Chalvon-Demersay, Sabine. 2012. "La part vivante des héros de série." In *Faire des sciences sociales. Critiquer*, edited by Pascale Haag and Cyril Lemieux, 31–57. Paris: Éditions de l'EHESS.

Champagne, Patrick. 1984. "La manifestation: La production de l'événement politique." *Actes de la recherche en sciences sociales* 52–53: 19–41.

Chompalov, Ivan, and Wesley Shrum. 1999. "Institutional Collaboration in Science: A Typology of Technological Practice." *Science, Technology, & Human Values* 24, no. 3: 338–372.

Clark, Candace. 1997. *Misery and Company: Sympathy in Everyday Life*. Chicago, IL: University of Chicago Press.

Clark, Colin, and Trevor Pinch. 1992. "The Anatomy of a Deception: Fraud and Finesse in the Mock Auction Sales 'Con.'" *Qualitative Sociology* 15: 151–175.

Clark, Colin, and Trevor Pinch. 1995. *The Hard Sell: The Language and Lessons of Street-Wise Marketing*. London: Harper Collins.

Cliquet, Étienne. 2010. *La démo*. http://www.ordigami.net/files/demo/.

Collins, Harry M. 1988. "Public Experiments and Displays of Virtuosity: The Core-Set Revisited." *Social Studies of Science* 18: 725–748.

Coopmans, Catelijne. 2011. "'Face Value': New Medical Imaging Software in Commercial View." *Social Studies of Science* 41: 155–156.

Cossart, Paula, and Emmanuel Taïeb. 2011. "Spectacle politique et participation. Entre médiatisation nécessaire et idéal de la citoyenneté." *Sociétés & Représentations* 31, no. 1: 137–156.

Crary, Jonathan. 2001. *Suspensions of Perception: Attention, Spectacle and Modern Culture*. Cambridge, MA: MIT Press.

Debord, Guy. (1967) 1994. *The Society of the Spectacle*. Translated by Donald Nicholson-Smith. New York: Zone Books.

De Carvalho, Lucie. 2018. "'Remember There's Nothing Secret About a Nuclear Power Station.' Institutional Communication on Invisible Environmental Risks in British TV Footage (1956–1982)." *Revue Française de Civilisation Britannique* 23, no. 3. http://journals.openedition.org/rfcb/2425.

Déchaux, Jean-Hugues. 2015. "Intégrer l'émotion à l'analyse sociologique de l'action." *Terrains/Théories* 2. http://journals.openedition.org/teth/208.

Dolza, Luisa, and Hélène Vérin. 2003. "Une mise en scène de la technique: Les théâtres de machines." *Alliage: Culture-Science-Technique* 50–51: 8–20.

Dulong, Renaud. 1998. *Le témoin oculaire: Les conditions sociales de l'attestation personnelle*. Paris: Éditions de l'EHESS.

Durkheim, Émile. (1897) 1951. *Suicide: A Study in Sociology*. Translated by John A. Spaulding and George Simpson. Glencoe, IL: The Free Press.

Durkheim, Émile. (1895) 1982. *The Rules of Sociological Method*. Translated by W. D. Halls. New York: The Free Press.

Duval, Maurice. 1981. "Les camelots." *Ethnologie française* 2: 145–170.

Edwards, Paul N. 1996. *The Closed World: Computers and the Politics of Discourse in Cold War America*. Cambridge, MA: MIT Press.

Elmer-DeWitt, Philip. 1988. "Technology: Soul of the Next Machine." *Time*, October 24, 80–82.

Ezrahi, Yaron. 1990. *The Descent of Icarus: Science and the Transformation of Contemporary Democracy*. Cambridge, MA: Harvard University Press.

Falk, Lorne, and Heidi Gilpin. 1995. "Demo Aesthetics." *Convergence: The Journal of Research into New Media Technologies* 2, no. 2: 127–139.

Favre, Pierre, ed. 1990. *La manifestation*. Paris: Presses de la FNSP.

Fillieule, Olivier. 1999. "'Plus ça change, moins ça change.' Demonstrations in France during the Nineteen-Eighties." In *Acts of Dissent: New Developments in the Study of Protest*, ed. Dieter Rucht, Ruud Koopmans, and Friedhelm Neidhardt, 199–226. Oxford: Rowman & Littlefield.

Fillieule, Olivier, and Fabien Jobard. 1998. "The Policing of Protest in France: Toward a Model of Protest Policing." In *Policing Protest: The Control of Mass Demonstrations in Western Democracies*, edited by Donatella della Porta and Herbert Reiter, 70–90. Minneapolis: University of Minnesota Press, 1998.

Fleck, Ludwig. 1979. *Genesis and Development of a Scientific Fact*. Chicago, IL: University of Chicago Press.

Frances, Jean. 2015. "Des doctorants en 'démo.' Ethnographie de deux sessions posters des Doctoriales." *Réseaux* 190–191, no. 2: 45–71.

Fraser, Nancy. 1992. "Rethinking the Public Sphere: A Contribution to the Critique of Actually Existing Democracy." In *Habermas and the Public Sphere*, edited by Craig Calhoun, 109–142. Cambridge, MA: MIT Press, 1992.

Gaglio, Gérald. 2009. "Faire des présentations, c'est travailler, mais comment? Diapositives numériques et chaîne d'activités dans des services marketing." *Activités* 6, no. 1: 111–138. https://journals.openedition.org/activites/2175.

Gallo, Carmine. 2010. *The Presentation Secrets of Steve Jobs: How to Be Insanely Great in Front of Any Audience*. New York: McGraw Hill.

Garfinkel, Harold. 1984. *Studies in Ethnomethodology*. Cambridge: Polity Press.

Geison, Gerald L. 1995. *The Private Science of Louis Pasteur*. Princeton, NJ: Princeton University Press.

Gingras, Yves. 2016. *Bibliometrics and Research Evaluation: Uses and Abuses*. Cambridge, MA: MIT Press.

Girard, Monique, and David Stark. 2007. "Socio-technologies of Assembly: Sense Making and Demonstration in Rebuilding Lower Manhattan." In *Governance and Information: The Rewiring of Governing and Deliberation in the 21st Century*, edited by David Lazer and Viktor Mayer-Schönberger, 145–176. Oxford: Oxford University Press.

Goffman, Erving. 1959. *The Presentation of Self in Everyday Life*. Garden City, NY: Doubleday.

Goffman, Erving. 1961. *Encounters: Two Studies in the Sociology of Interaction*. Indianapolis, IN: Bobbs-Merrill.

Goffman, Erving. 1963. *Behavior in Public Places: Note on the Social Organization of Gatherings*. New York: The Free Press.

Goffman, Erving. 1967. *Interaction Ritual: Essays on Face-to-Face Behavior*. Garden City, NY: Anchor Books.

Goffman, Erving. 1974. *Frame Analysis: An Essay on the Organization of Experience*. Cambridge, MA: Harvard University Press.

Goldstein, Catherine. 1995. *Un théorème de Fermat et ses lecteurs*. Saint Denis: Presses universitaires de Vincennes.

Goscinny, René and Albert Uderzo. (1972) 2004. *Asterix and the Soothsayer*. Translated by Anthea Bell and Derek Hockridge. London: Orion.

Greene, Jeremy A. 2004. "Etiquette and the Pharmaceutical Salesman in Postwar America." *Social Studies of Science* 34, no. 2: 271–292.

Grelon, André, Françoise Chamozzi, and Ina Wagner. 2000. "De la technique comme spectacle." *Alliage* 50–51: 2–5.

Grimaud, Emmanuel. 2003. *Bollywood Film Studio, ou comment les films se font à Bombay*. Paris: CNRS Éditions.

Grimaud, Emmanuel. 2007. "The Film in Hand: Modes of Coordination and Assisted Virtuosity in the Bombay Film Studios." *Qualitative Sociology Review* 3, no. 3: 59–77.

Hankins, Thomas L., and Robert J. Silverman. 1995. *Instruments and the Imagination*. Princeton, NJ: Princeton University Press.

Hilgartner, Stephen. 2000. *Science on Stage: Expert Advice as Public Drama*. Stanford, CA: Stanford University Press.

Hindle, Jim. 2006. *Nine Miles: Two Winters of Anti-Road Protest*. London: Underhill Books.

Hochschild, Arlie R. 1983. *The Managed Heart: Commercialization of Human Feeling*. Berkeley, CA: University of California Press.

Houdart, Sophie. 2005. "'Ceci est une persienne.' Montrer/démontrer dans un projet architectural au Japon." Presented at the Sociology of Demonstration Workshop, May 19–20. Marseille: CNRS.

Houellebecq, Michel. 2011. *Whatever*. Translated by Paul Hammond. London: Serpent's Tail.

Hubert, Matthieu, and Dominique Vinck. 2017. *Nanotechnologies: L'invisible révolution*. Paris: Le Cavalier Bleu.

Jaffré, Maxime. 2005. "La décontextualisation des musiques traditionnelles: Le cas de la musique arobo-andalouse à Marseille." MA thesis. Marseille: EHESS.

Jasanoff, Sheila. 2005. *Designs on Nature: Science and Democracy in Europe and the United States*. Princeton, NJ: Princeton University Press.

Jones, Graham. 2011. *Trade of the Tricks: Inside the Magician's Craft*. Berkeley, CA: University of California Press.

Jouvenet, Morgan, Jérôme Lamy, and Arnaud Saint-Martin. 2015. "Les activités spatiales, objet sociologique." *Histoire de la recherche contemporaine* 4, no. 2: 171–179.

Kalberg, Stephen. 2012. *Max Weber's Comparative-Historical Sociology Today: Major Themes, Modes of Analysis, and Applications*. London: Routledge.

Karpik, Lucien. 1995. *Les avocats: Entre l'État, le public et le marché, XIIIe–XXe siècles*. Paris: Gallimard.

Kawasaki, Guy, and Michele Moreno. 2000. *Rules For Revolutionaries: The Capitalist Manifesto for Creating and Marketing New Products and Services*. New York: Harper Business.

Keith, William. 2007. *Democracy as Discussion: Civic Education and the American Forum Movement*. Lanham, MD: Lexington Books.

Kuhn, Thomas. 1962. *The Structure of Scientific Revolutions*. Chicago, IL: University of Chicago Press.

LaGrandeur, Kevin. 2003. "Digital Images and Classical Persuasion." In *Eloquent Images: Word and Image in the Age of New Media*, edited by Mary E. Hocks and Michelle R. Kendrick, 117–136. Cambridge, MA: MIT Press.

Lahire, Bernard. 2013. *Dans les plis singuliers du social: Individus, institutions, socialisations*. Paris: La Découverte.

Lakatos, Imre. 1976. *Proofs and Refutations: The Logic of Mathematical Discovery*. Edited by John Worrall and Elie Zahar. Cambridge, MA: Cambridge University Press.

Lamont, Michèle. 2009. *How Professors Think: Inside the Curious World of Academic Judgment*. Cambridge, MA: Harvard University Press.

Lampel, Joseph. 2001. "Show-and-Tell: Product Demonstrations and Path Creation of Technological Change." In *Path Dependence and Creation*, edited by Raghu Garud and Peter Karnoe, 303–328. Mahwah, NJ: Lawrence Erlbaum.

Lamy, Jérôme. 2017. "La science à la cour de Versailles: Mise en scène du savoir et démonstration du pouvoir (XVIIe-XVIIIe siècles)." *Cahiers d'histoire. Revue d'histoire critique* 136: 71–99.

Latour, Bruno. 1983. "Give Me a Laboratory and I Will Raise the World." In *Science Observed*, edited by Karin Knorr-Cetina and Michael Mulkay, 141–170. Beverly Hills, CA: Sage.

Laurent, Brice. 2011. "Technologies of Democracy: Experiments and Demonstrations." *Science & Engineering Ethics* 17, no. 4: 649–666.

Laurent, Brice. 2017. *Democratic Experiments: Problematizing Nanotechnology and Democracy in Europe and the United States*. Cambridge, MA: MIT Press.

Le Bon, Gustave. (1895) 2002. *The Crowd: A Study of the Popular Mind*. Mineola, NY: Dover Publications.

Le Velly, Ronan. 2007. "Les démonstrateurs de foires. Des professionnels de l'interaction symbolique." *Ethnologie française* 37: 143–151.

Lehman, Christine. 2010. "Les miroirs ardents: Recherches académiques et divertissements à la cour." In *Sciences et curiosités à la cour de Versailles*, edited by Béatrix Saule and Catherine Arminjon, 234–236. Coédition Rmn éditions / Établissement public du musée et du domaine de Versailles.

Lemieux, Cyril. 2008. "Rendre visibles les dangers du nucléaire: Une contribution à la sociologie de la mobilisation." In *La cognition au prisme des sciences sociales*, edited by Bernard Lahire and Claude Rosental, 131–159. Paris: Éditions des Archives Contemporaines.

Leroy, Marie-Laure. 2008. "Jeremy Bentham ou la sympathie pour le plus grand nombre." *Revue du Mauss* 31: 122–136.

Licoppe, Christian. 1996. *La formation de la pratique scientifique: Le discours de l'expérience en France et en Angleterre (1630–1820)*. Paris: La Découverte.

Lloyd, Geoffrey. 1990. "The Theories and Practices of Demonstration in Aristotle." *Proceedings of the Boston Area Colloquium in Ancient Philosophy* 6, no. 1: 371–401.

Lunenfeld, Peter. 2000. *Snap to Grid: A User's Guide to Digital Arts, Media, and Cultures*. Cambridge, MA: MIT Press.

Lunenfeld, Peter. 2016. "Demo n.0." In *Practicable: From Participation to Interaction in Contemporary Art*, edited by Samuel Bianchini and Erik Verhagen, 345–357. Cambridge, MA: MIT Press.

Lynch, Michael E., and David Bogen. 1996. *The Spectacle of History: Speech, Text, and Memory at the Iran-Contra Hearings*. Durham, NC: Duke University Press.

MacKenzie, Donald. 1990. *Inventing Accuracy: A Historical Sociology of Nuclear Missile Guidance*. Cambridge, MA: MIT Press.

MacKenzie, Donald. 1994. "Computer-Related Accidental Death: An Empirical Exploration." *Science and Public Policy* 21, no. 4: 233–248.

MacKenzie, Donald. 1996. "Safety-critical and Security-critical Computing in Britain: An Exploration." *Technology Analysis & Strategic Management* 8, no. 4: 355–379.

Mariot, Nicolas. 2001. "Les formes élémentaires de l'effervescence collective, ou l'état d'esprit prêté aux foules." *Revue française de science politique* 5: 707–738.

Markoff, John. 1996. "Nothing Up Their Sleeves? Masters of the High-tech Demo Spin Their Magic." *New York Times*, March 11, D1.

Markusson, Nils, Atsushi Ishii, and Jennie C. Stephens. 2011. "The Social and Political Complexities of Learning in Carbon Capture and Storage Demonstration Projects." *Global Environmental Change* 21, no. 2: 293–302.

Mauss, Marcel. 1921. "L'expression obligatoire des sentiments (rituels oraux funéraires australiens)." *Journal de Psychologie* 18: 425–434.

Mauss, Marcel. (1925) 1990. *The Gift: The Form and Reason for Exchange in Archaic Societies*. Translated by W. D. Halls. London: Routledge.

McCurdy, Howard E. 2001. *Faster, Better, Cheaper: Low-Cost Innovation in the U.S. Space Program*. Baltimore, ME: Johns Hopkins University Press.

Merton, Robert K. 1938. "Science and the Social Order." *Philosophy of Science* 5: 321–337.

Mischi, Julian. 2013. "Savoirs militants et rapports aux intellectuels dans un syndicat cheminot." *Actes de la recherche en sciences sociales* 196–197: 132–151.

Molinié, Antoinette. 2016. *La Passion selon Séville*. Paris: CNRS Éditions.

Mukerji, Chandra. 1997. *Territorial Ambitions and the Gardens of Versailles.* Cambridge, MA: Cambridge University Press.

Mukerji, Chandra. 2009. *Impossible Engineering: Technology and Territoriality on the Canal du Midi.* Princeton, NJ: Princeton University Press.

O'Neill, Rebeca Neri, and Alain Nadaï. 2012. "Risque et démonstration: La politique de capture et de stockage du dioxyde de carbone (CCS) dans l'Union européenne." *VertigO—la revue électronique en sciences de l'environnement* 12, no. 1. http://journals.openedition.org/vertigo/12172.

Pfaffenberger, Bryan. 1992. "Technological Dramas." *Science, Technology and Human Values* 17, no. 3: 282–312.

Pinch, Trevor. 1993. "Testing—One, Two, Three . . . Testing! Toward a Sociology of Testing." *Science, Technology and Human Values* 18, no. 1: 25–41.

Pinch, Trevor. 2003. "Giving Birth to New Users: How the Minimoog Was Sold to Rock and Roll." In *How Users Matter: The Co-Construction of Users and Technology,* edited by Nelly Oudshoorn and Trevor Pinch, 247–270. Cambridge, MA: MIT Press.

Pinch, Trevor, and Frank Trocco. 2002. *Analog Days: The Invention and Impact of the Moog Synthesizer.* Cambridge, MA: Harvard University Press.

Pollock, Neil, and Robin Williams. 2010. "The Business of Expectations: How Promissory Organizations Shape Technology and Innovation." *Social Studies of Science* 40, no. 4: 525–548.

Pratt, Vernon. 1987. *Thinking Machines: The Evolution of Artificial Intelligence.* Oxford: Blackwell.

Proust, Marcel. (1923) 1927. *La Captive.* In *Remembrance of Things Past.* Translated by C. K. Scott Moncrieff, vol. 2, 383–669. New York: Random House.

Quéré, Louis. 1992. "L'espace public: De la théorie politique à la métathéorie sociologique." *Quaderni* 18: 75–92.

Quéré, Louis. 2008. "Les neurosciences fournissent-elles une explication 'plus' scientifique des phénomènes socio-culturels?" In *La cognition au prisme des sciences sociales,* edited by Bernard Lahire and Claude Rosental, 23–54. Paris: Éditions des Archives Contemporaines.

Rosental, Claude. 2000. "Les travailleurs de la preuve sur Internet: Transformations et permanences du fonctionnement de la recherche." *Actes de la recherche en sciences sociales* 134: 37–44.

Rosental, Claude. 2002. "De la démo-cratie en Amérique: Formes actuelles de la démonstration en intelligence artificielle." *Actes de la recherche en sciences sociales,* 141–142: 110–120.

Rosental, Claude. 2003. "Certifying Knowledge: The Sociology of a Logical Theorem in Artificial Intelligence." *American Sociological Review* 68: 623–644.

Rosental, Claude. 2004. "Fuzzyfying the World: Social Practices of Showing the Properties of Fuzzy Logic." In *Growing Explanations: Historical Perspectives on Recent Science*, edited by M. Norton Wise, 159–178. Durham, NC: Duke University Press.

Rosental, Claude. 2005. "Making Science and Technology Results Public: A Sociology of Demos." In *Making Things Public: Atmospheres of Democracy*, edited by Bruno Latour and Peter Weibel, 346–349. Cambridge, MA: MIT Press.

Rosental, Claude. 2007. *Les capitalistes de la science: Enquête sur les démonstrateurs de la Silicon Valley et de la NASA*. Paris: CNRS Éditions.

Rosental, Claude. 2008. *Weaving Self-Evidence: A Sociology of Logic*. Translated by Catherine Porter. Princeton, NJ: Princeton University Press.

Rosental, Claude. 2009. "Anthropologie de la démonstration." *Revue d'Anthropologie des Connaissances* 3: 233–252.

Rosental, Claude. 2010. "Social Studies of Evaluation." *Social Studies of Science* 40: 481–484.

Rosental, Claude. 2011. "Eco-Demos: Using Public Demonstrations to Influence and Manage Environmental Choices and Politics." *Institut Marcel Mauss–CEMS Occasional Paper N°3*. http://cems.ehess.fr/index.php?2671.

Rosental, Claude. 2013. "Toward a Sociology of Public Demonstrations." *Sociological Theory* 31, no. 4: 343–365.

Rosental, Claude. 2015. "From Numbers to Demos: Assessing, Managing and Advertising European Research." *Histoire de la recherche contemporaine* 4, no. 2: 163–170.

Rosental, Claude. 2017. "Modes of Exchange: The Cultures and Politics of Public Demonstrations." In *Cultures without Culturalism: The Making of Scientific Knowledge*, edited by Karine Chemla and Evelyn Fox Keller, 170–195. Durham, NC: Duke University Press.

Rosental, Claude. 2018. "Entre prouesses verbales et virtuosité technique. Les démonstrations publiques de technologie." In *Les objets composés*, edited by Nicolas Dodier and Anthony Stavrianakis, 237–266. Paris: Éditions de l'EHESS.

Rosental, Claude. 2019. "Formuler des promesses technologiques à l'aide de demos." *Socio* 12: 27–47.

Russell, Stewart, Nils Markusson, and Vivian Scott. 2012. "What Will CCS Demonstrations Demonstrate?" *Mitigation and Adaptation Strategies for Global Change* 17, no. 6: 651–668.

Sacks, Harvey, Emmanuel A. Schegloff, and Gail Jefferson. 1974. "A Simplest Systematics for the Organization of Turn-Taking for Conversation." *Language* 50, no. 4: 696–735.

Schaffer, Simon. 1994. "Machine Philosophy: Demonstration Devices in Georgian Mechanics." *Osiris* 9: 157–182.

Shapin, Steven. 1984. "Pump and Circumstance: Robert Boyle's Literary Technology." *Social Studies of Science* 14, no. 4: 481–520.

Shapin, Steven. 1988. "The House of Experiment in Seventeenth Century England." *Isis* 79: 373–404.

Shapin, Steven, and Simon Schaffer. 1985. *Leviathan and the Air-Pump: Hobbes, Boyle, and the Experimental Life*. Princeton, NJ: Princeton University Press.

Shepard, Benjamin. 2009. *Queer Political Performance and Protest: Play, Pleasure, and Social Movement*. London: Routledge.

Sherry, John F. 1998. "Market Pitching and the Ethnography of Speaking." *Advances in Consumer Research* 15: 543–547.

Simakova, Elena. 2010. "RFID 'Theatre of the Proof': Product Launch and Technology Demonstration as Corporate Practices." *Social Studies of Science* 40: 549–576.

Smith, Wally. 2009. "Theatre of Use: A Frame Analysis of IT Demonstrations." *Social Studies of Science* 39: 449–480.

Stark, David, and Verena Paravel. 2008. "Powerpoint in Public: Digital Technologies and the New Morphology of Demonstration." *Theory, Culture and Society* 25: 30–55.

Sutcliffe, Mark. 2013. "Supernova? No, Supermarketer: Hadfield Offered Classic Lessons in Marketing Any Product." *The Ottawa Citizen*, May 17, F1.

Tarde, Gabriel de. (1890) 1903. *The Laws of Imitation*. Translated by Elsie Clews Parsons. New York: Henry Holt.

Tardy, Cécile, and Yves Jeanneret. 2006. "'Profondeurs de l'urgent': *PowerPoint*, entre immédiateté et mémoire." *Communication & organisation* 29: 167–179.

Tarrow, Sidney. 1994. *Power in Movement: Collective Action, Social Movements and Politics*. Cambridge: Cambridge University Press.

Thébaud-Sorger, Marie. 2009. *L'Aérostation au temps des Lumières*. Rennes: Presses universitaires de Rennes.

Thénard, Jean-Michel. 2015. "Royal ne veut pas être traînée dans la boue (rouge)." *Canard Enchaîné*, May 27, 3.

Tilly, Charles. 1986. *The Contentious French*. Cambridge, MA: Harvard University Press.

Tocqueville, Alexis de. 2000. *Democracy in America*. Translated by Harvey C. Mansfield and Delba Winthrop. Chicago: University of Chicago Press.

Topçu, Sezin. 2013. *La France nucléaire: L'art de gouverner une technologie contestée*. Paris: Seuil.

Tournay, Virginie. 2014. *Penser le changement institutionnel: Essai sur la logique évolutionnaire*. Paris: Presses universitaires de France.

Traïni, Christophe. 2010. "Dramaturgie des émotions, traces des sensibilités. Observer et comprendre des manifestations anti-corrida." *ethnographiques.org* 21. http://www.ethnographiques.org/2010/Traini.

Turner, Jonathan H., and Jan E. Stets. 2006. "Sociological Theories of Human Emotions." *Annual Review of Sociology* 32: 25–52.

Van Dijck, José. 2001. "Bodyworlds: The Art of Plastinated Cadavers." *Configurations* 9, no. 1: 99–126.

Vargha, Zsuzsanna. 2011. "From Long-Term Savings to Instant Mortgages: Financial Demonstrations and the Role of Interaction in Markets." *Organization* 18, no. 2: 215–235.

Vaughan, Diane. 1996. *The Challenger Launch Decision: Risky Technology, Culture, and Deviance at NASA*. Chicago, IL: University of Chicago Press.

Vidal, Denis. 2006. "Les sirènes de l'expérience. Populisme expérimental ou démocratie du jugement." *Terrain* 46: 67–84.

Vinck, Dominique. 2009a. "De l'objet intermédiaire à l'objet-frontière: Vers la prise en compte du travail d'équipement." *Revue d'Anthropologie des Connaissances* 3, no. 1: 51–72.

Vinck, Dominique. 2009b. *Les nanotechnologies*. Paris: Le Cavalier Bleu.

Vinck, Dominique. 2011. "Taking Intermediary Objects and Equipping Work into Account in the Study of Engineering Practices." *Engineering Studies* 3, no. 1: 25–44.

Vom Lehn, Dirk, Christian Heath, and Jon Hindmarsh. 2001. "Exhibiting Interaction: Conduct and Collaboration in Museums and Galleries." *Symbolic Interaction* 24, no. 2: 189–216.

Von Staden, Heinrich. 1994. "Anatomy as Rhetoric: Galen on Dissection and Persuasion." *Journal of the History of Medicine and Allied Sciences* 50: 47–66.

Wagner, Annette, and Maria Capucciati. 1996. "Demo or Die: User Interface as Marketing Theatre." *CHI–96 Electronic Proceedings*. dl.acm.org/citation.cfm?doid=238386.238606.

Werrett, Simon. 2009. "William Congreve's Rational Rockets." *Notes & Records of the Royal Society* 63: 35–56.

Winthereik, Brit Ross, Nis Johannsen, and Dixi Louise Strand. 2008. "Making Technology Public: Challenging the Notion of Script through an E-health Demonstration Video." *Information, Technology and People* 21: 116–132.

Woolgar, Steve. 1991. "Configuring the User: The Case of Usability Trials." In *A Sociology of Monsters: Essays on Power, Technology and Domination*, edited by John Law, 58–99. London: Routledge.

Yaneva, Albena. 2009. "The Architectural Presentation: Techniques and Politics." In *Networks of Design*, edited by Jonathan Glynne, Fiona Hackney, and Viv Minton, 212–219. Boca Raton, FL: Universal Publishers.

Yates, JoAnne, and Wanda Orlikowski. 2007. "The Power Point Presentation and Its Corollaries: How Genres Shape Communicative Action in Organizations." In *Communicative Practices in Workplaces and the Professions: Cultural Perspectives on the Regulation of Discourse and Organizations*, edited by Marc Zachary and Charlotte Thralls, 67–91. New York: Baywood.

INDEX

Ability
 of audiences, 22, 35, 119, 137
 of demonstrators, 70, 72, 133, 161, 170, 179, 180, 182, 185, 196, 198, 200 (*see also* Competency: of demonstrators)
Acclamation, 29, 75, 231
Accompaniment of demonstrations, 43, 73, 77, 80, 81, 83, 105, 115, 166, 188, 204, 218, 219
ACTS (Advanced Communication Technologies and Services), 119–123, 125, 153
Adaptation
 of demonstrations, 11, 111, 135, 143, 170, 179, 212, 229, 234
 of a performance to an audience, 115–117, 131, 135, 142, 148, 169, 171, 177 (*see also* Adjustment of a performance to an audience)
Adjustment of a performance to an audience, 17, 67, 82, 93, 109, 117, 131, 143, 198, 234. *See also* Adaptation: of a performance to an audience

Admiration, 82, 91, 126, 128, 203, 204
Advertisements, 20, 52, 68, 100, 115, 136, 139, 168, 172, 180, 188, 212. *See also* Marketing; Promotion of products
Advice for the development of demonstrations. *See also* Prescriptions for the development of demonstrations
 from communications consultants, 47, 49, 76, 79, 158, 160, 165, 167, 171
 from demonstrators, 60, 93, 169, 170, 183, 185, 186, 191, 205, 207, 209
 from police officers, 172, 173
 from researchers, 50, 168
 from technology evangelists, x, 3, 46, 82, 157, 158, 172
Ambiguity, 23, 24, 36, 154, 232
Ambivalence, 18, 152, 154, 190, 229
Amusement of audiences, 44, 69, 70, 115, 147, 148, 150, 196, 199, 201, 206–208
Anatomy lessons, 4, 175, 176, 212, 215, 216. *See also* Dissection

Anecdotes, 47, 80, 158, 207, 210
Animation within demonstrations, 55, 57, 58, 65, 66, 88, 92, 94, 110, 122, 123, 221
Animators, 181, 182
Anticipation of audiences' reactions, 52, 79, 81, 134, 228
Apparatus, 41, 42, 63, 81, 87, 127, 128, 136, 162, 214, 220, 225, 228, 229, 248n1
Appeal of demonstrations, 54, 121, 204
Appreciation. *See also* Assessment; Evaluation; Judgments
  of demonstrations, 76, 136, 138, 147, 194, 203, 207–210
  of technologies, 42, 71, 107, 137
Apprenticeship, 166–168, 171, 173, 191. *See also* Transmission of demonstrative skills
Architecture, 8, 15, 37, 65, 66, 114
Arenas, 15, 87, 112, 175. *See also* Forums
Arguments, 11, 54, 56, 59, 63, 65, 73, 108, 114–116, 126, 137, 142, 152, 153, 165, 170, 179
  verbal, 61–63, 66–69, 74, 116, 117, 188
Arrangement, 23, 89, 101, 128, 157, 160
  of material nature, 17, 58, 218
  of spaces, 39, 65, 91, 95, 103
Artificial intelligence, 2, 11, 23, 28, 49, 62, 100, 111, 113, 130, 143, 167, 168, 170, 171, 174, 175
Assessment. *See also* Appreciation; Evaluation; Judgments
  of demonstrations, 30, 164
  of policies, 137
  of technologies, 71
Astonishment of audiences, 24, 44, 96, 144, 202, 213. *See also* Surprises
Attention, 43, 44, 63–65, 80, 89, 90, 92, 93, 95, 110, 141, 183, 189, 216
  of spectators, 24, 26, 41, 42, 49, 51, 66, 67, 69, 76, 121, 158, 231

Attraction of audiences, 35, 59, 97, 121, 140, 153, 170, 181, 182, 188, 189, 191, 194, 197, 198, 200–202, 206, 212, 213, 225
Audiard, Michel, 73, 156
Audience engagement. *See* Emotions; Engagement: of audiences; Feelings; Play on affects
Authenticity, 76, 175
Authorities, x, 102, 115, 119, 120, 125, 129, 130, 163, 164, 187, 189, 199, 235
Awareness, v, 31
  of audiences, 26, 182, 184
  of demonstrators, 107, 156, 169

Barkers, 61, 75, 189. *See also* Hucksters; Peddlers; Posticheurs
Barnum, Phineas, 152, 153, 193, 213
Basketonics, 53–59, 69, 94
Bienvenu, François, 193, 212
Body engagement, 55, 56, 70, 86, 130, 185, 216, 224, 225
Booth, 81–83, 85, 86, 95, 102, 112
Brand ambassadors, 181–183, 187. *See also* Business representatives; Sales representatives
Bugs, 18, 39, 51, 80–82, 90, 146, 148, 150, 154, 156–161, 166, 203, 230. *See also* Crashes; Flaws; Malfunctions
Business representatives, 28, 39, 138, 234. *See also* Brand ambassadors; Sales representatives
Buyer, 8, 10, 22, 47, 61, 102, 206, 214

Calibration of demonstrations, 45, 49, 61, 76, 228
Capitalism, critique of, 18, 151, 156, 216, 217, 248n1
Capitalization of demonstrative efforts, 110, 111, 118
Ceremonials, x, 18, 21, 35
Chain of actions, 102, 131, 235

INDEX

Charisma, 74, 176
Charm, 70, 71. *See also* Seduction
Charts, 54–57, 94, 120, 122, 125, 129, 134. *See also* Graphics; Visual effects
Clothing, 65, 174, 178. *See also* Costumes
Collaboration, 42, 49, 90, 95, 116, 217, 222, 240n5
Combination
  of demonstrations, 11, 114, 115, 127, 131, 133, 235
  of demonstrative elements, 66, 68, 92, 101, 117, 138, 148, 210, 212, 219, 222, 227
Commentary
  on demonstrations, 16, 62, 71, 100, 148, 150, 170, 204, 206–211, 245n6, 245n14
  by demonstrators, 42, 54, 67, 83, 113, 115, 135, 137, 145, 151, 169, 170, 198, 204, 206–208, 210, 218
Commercial spectacle, x, 18, 199, 202, 233
Communications consultant, 45–47, 49, 51, 70, 76, 96, 158, 171, 176
Comparison of products, 46, 48, 53, 54, 58–60, 76, 127
Competency
  of demonstratees, 142, 152, 159, 231
  demonstration of, 7, 180, 236
  of demonstrators, 53, 95, 96, 166, 173, 177, 179, 180, 231 (*see also* Ability: of demonstrators)
Competition, 18, 49, 52, 66, 76, 82, 85, 88, 113, 119, 124, 141, 142, 152, 163, 164, 189, 200, 204, 220, 229, 245n6
Complicity of audiences, 17, 24, 25, 41, 45, 51, 60, 69, 71, 72, 148, 161
Composition
  of audiences, 18, 132, 140–142, 165, 229, 231
  of demonstrations, 217

Concreteness of demonstrations, 81, 87, 94, 109, 120. *See also* Tangibility of demonstrations
Confidence, 24, 136–138, 163, 213
Confrontation of demonstrations, 11, 63, 114, 124, 126, 130, 131, 163, 171, 228
Consensus, 13, 126, 138, 199
Control
  of crowds, 18, 20, 34, 43, 90, 92, 126, 129, 135, 140, 141, 217
  of interpretations, 10, 17, 81, 143, 155, 227, 234
Conventions, 17, 20, 24, 25, 178, 231, 232, 239n10
Conversion of users, 46, 74, 237n4
Conviction of audiences, 25, 32, 65, 102, 115, 118, 149, 152, 170, 233
Coordination of demonstrators, 89, 103, 105, 108, 111, 112, 127, 228, 229
Costumes, 29, 64, 103, 172, 175, 200. *See also* Clothing
Crashes, 25, 81, 82, 90, 156, 159, 230. *See also* Bugs; Flaws; Malfunctions
Credibility, 25, 34, 53, 68, 87, 108, 112, 131, 135, 138, 158, 170, 199
Critical habits, 17, 118
Criticism on the part of audiences, 109, 119, 120, 171
Cumulative effects of demonstrations, 107, 113, 131
Curiosity of audiences, 44, 72, 97–99, 153, 195, 197, 199, 213–215

Décor, 24, 92, 95, 103, 145, 208
Delta project, 51, 80, 81, 93, 94, 99, 105–108, 110, 111, 115–119, 135, 136, 157, 174
"Demo god," x, 3, 10, 46, 187
Demo mode, 8, 223. See also *Mode: demo*
Demonstrations of strength, 7, 34, 91, 92, 101, 108, 113, 114, 123, 131

Demo or die, 2, 39, 174, 218, 224, 232
Detail man, 177–180, 185
Dialectics, 143, 171, 232
Didactics, 72, 239n11. *See also* Pedagogy; Teaching with demonstrations
Disappointment, 98, 99
Discourse, 6, 41, 45, 58, 61, 63, 66–69, 76, 125, 128, 136, 202, 218, 219, 222. *See also* Speeches; Talks; Verbal performances
Dissection, 4, 215. *See also* Anatomy lessons
Doubt, 28, 52, 93, 117, 130, 153, 155, 213. *See also* Skepticism
Dramaturgy, 10, 16, 18, 24, 70, 76, 80, 240n8
Dreams of spectators, 52, 70. *See also* Imagination of spectators
Duration of performances, 100, 105, 116. *See also* Sequences; Temporality; Timing
Durkheim, Emile, 61, 75

EDF, 126–129
Edison, Thomas, 25, 98, 143
Effectiveness of demonstrations, 17, 53, 68, 73, 93, 103, 112, 114, 115, 128, 131, 155, 166, 169. *See also* Efficiency of demonstrations
Efficiency of demonstrations, 41, 94, 161. *See also* Effectiveness of demonstrations
Elkan, Charles, 62, 63, 112, 151, 152
Emotions. *See also* Feelings; Play on affects
　looking for, 29
　manifestation of, 4, 7, 208
　play on, 17, 24, 36, 37, 47, 68, 70–77, 99, 142, 227
Empathy, 69, 137. *See also* Sympathy
Energy of demonstrators, 16, 40, 71, 74, 79, 80, 203, 233

Engagement
　of audiences, 71, 75, 213
　of demonstrators with the public, 29, 38, 47, 182
Enjoyment, 50, 55, 71, 194, 195, 210
Enthusiasm. *See also* Euphoria; Passion
　of audiences, 69, 70, 98–101, 117, 118, 121, 123, 133, 149, 150, 155, 164, 205
　of demonstrators, 38, 48, 58, 72–75, 86, 184, 187
Entrepreneurs, 3, 111, 174, 198
Equipment. *See* Verification of equipment
Ethnomethodology, 86, 160, 240n3
Euphoria, 71, 99, 199. *See also* Enthusiasm; Passion
European Commission, x, 2, 120, 121, 123
Evaluation. *See also* Appreciation; Assessment; Judgments
　of demonstrations, 231
　of demonstrators, 96
　of technologies, 24, 32, 71, 99, 109, 116, 120, 122, 137, 206, 210
Exaggeration, 75, 135, 175
Exceptionality of what is on display, 8, 58, 59, 61, 122, 198. *See also* Uniqueness: of what is on display
Exhibitions, 7, 8, 15, 50, 65, 73, 94, 95, 102, 106, 107, 112, 142, 152, 153, 156, 173, 198, 202, 205, 213–215, 218–220, 222–224, 234, 238n12
Expertise
　in communication, 45, 49, 51, 70
　of demonstrators, 18, 61, 77, 108, 198
Explanations, 46, 47, 57, 58, 82, 83, 95, 127, 188, 199, 219
Exploitation of demonstrations, 133, 193, 211, 213–215, 219, 225
Exploits, 91, 204. *See also* Prowess; Virtuosity

INDEX

Fabrication, 22, 35. *See also* Fakery; Fiction
Failure of demonstrations, 18, 24, 82, 132, 139, 151, 154, 157, 158, 160–165, 229
Faith, x, 3, 7, 29, 30, 234, 237n4
Fakery, 129, 153, 213. *See also* Fabrication; Fiction
Familiarity
 of demonstrators, 48, 86, 148
 produced by demonstrations, 127, 128, 142, 199
Fascination, 115, 121, 212, 214
Feasibility of an approach, 38, 50, 66, 120, 124, 179, 225
Feedback, 87, 181, 186
Feelings. *See also* Emotions: play on; Play on affects
 expressed by audiences, 210
 expressed by demonstrators, 137, 195, 236
 generated by demonstrators, 71, 72
Festiveness, 64, 69, 199, 200
Fiction, 17, 19, 22–27, 35, 89, 156, 180, 227, 231, 232. *See also* Fabrication; Fakery
Flaws, 51, 81, 114, 161, 205. *See also* Bugs; Crashes; Malfunctions
Format, 3, 38, 116–118, 122, 244n3
Forums, 62, 63, 111, 112, 151, 200, 244n3. *See also* Arenas
Fuzzy Logic, 62, 111, 112, 151, 170, 171

Gallo, Carmine, 46–49, 54, 70, 71, 74, 79, 95, 118, 158, 160, 171, 239n7
Gatherings, 4, 5, 7, 12, 32, 43, 83, 89, 91, 92, 140, 194, 200, 202, 211, 220, 221, 225, 235
Gaze, 50, 219
Gestures of demonstrators, 17, 61, 68, 70, 71, 77, 79, 96, 137, 172, 179, 207, 208, 210, 219, 222–224, 227

Goffman, Erving, 6, 16, 18–23, 28, 29, 31, 32, 35–37, 72, 128, 231, 237n1 (chap. 1)
Graphics, 24, 52, 54, 58, 66, 90, 94, 120, 125, 134, 135, 203, 217, 221, 222. *See also* Charts; Visual effects
Greenpeace, 10, 12, 126, 131, 134. *See also* Nongovernmental organizations

Hackers, 15, 70, 82, 151, 156
Highlights
 of competencies, 198, 203, 209, 213, 225
 of technological features, 24, 26, 67, 86, 87, 93, 122, 161
Hot-air balloons, 99, 193, 198, 211
Hucksters, 25, 188, 190, 244n31. *See also* Barkers; Peddlers; Posticheurs
Humor of demonstrators, 41, 43, 45, 48, 51, 60, 69, 71, 75, 86, 92, 129, 148, 151, 160, 161, 165, 175, 178, 184, 191, 200, 205–207, 219. *See also* Irony; Jokes of demonstrators

Ideal audience, 32, 34, 35, 41, 45, 229
Identity
 of audiences, 34–35, 134, 140, 142, 149, 152, 155, 163, 231
 of demonstrators, 65, 142, 180
Illusion, 25, 36. *See also* Manipulation: of audiences; Mystification
Imagination of spectators, 24, 71. *See also* Dreams of spectators
Impatience, 98, 100, 202
Imperative
 to demonstrate, x, 2, 174, 218, 233
 emotional, 72, 73, 77
Impressive demonstrations, 4, 7, 26, 47, 70, 75, 108, 123, 143, 144, 148, 150, 165, 182, 198, 214
Improvisation, 80, 82, 83, 96, 102, 156
Influencing audiences, 10, 25, 63, 71, 134, 154, 233

Interchangeable individuals, audience as, 28, 31, 36, 73, 140, 231
Interpretation of demonstrations, 10, 17, 41, 44, 101–103, 112, 126, 139, 152, 169, 227, 228, 234
Invisibility, 42, 51, 93
Irony. *See also* Humor of demonstrators; Jokes of demonstrators
  of audiences, 75
  of demonstrators, 27, 75, 175, 216, 217, 219, 224, 226, 233
Iterations in the preparation for demonstrations, 17, 86–88, 103, 109, 229

Jobs, Steve, 1, 8, 9, 24, 25, 31, 37, 46–49, 53, 54, 70, 71, 74, 79, 80, 86, 95, 98, 133, 141, 142, 157, 161, 164, 168, 171, 174, 175, 185, 234, 237n4, 239n7
Jokes of demonstrators, 27, 76, 148, 161, 194. *See also* Humor of demonstrators; Irony
Judgments. *See also* Appreciation; Assessment; Evaluation
  on demonstrations, 49, 119, 139, 151–153, 162, 163, 165, 166, 229
  on technologies, 10

Kawasaki, Guy, 45–47, 54, 82, 148, 157, 158, 171, 176, 187

Laughter. *See also* Smiling, as a need for demonstrators
  in the audience, 24, 25, 71, 150, 191, 206
  of demonstrators, 146, 159, 160, 196, 197
Launch of products, x, 8, 21, 23–25, 31, 38, 47, 48, 68, 71, 98, 100, 112, 126, 141, 157, 161, 176, 181, 198, 206
Le Bon, Gustave, 74, 139
Lectures, 4, 82, 112, 125, 218, 219

Legitimacy, 69, 102, 121, 138, 172, 177
Legitimization, 23, 111, 129, 138

Magicians, 24, 36, 51, 172. *See also* Prestidigitation
Makeup, ix, 207–210
Malfunctions, 134, 156, 157, 160–162, 165. *See also* Bugs; Crashes; Flaws
Manifestation
  of aptitudes, 169
  of the divine, 9
  of faith, intentions, inclinations, or emotion, 4, 7, 141, 175, 234
  as protest, 5, 32
Manipulation
  of audiences, 22, 36, 129, 139, 231, 232, 245n6, 248n1 (*see also* Illusion; Mystification)
  of technologies, 26, 30, 42, 51, 57, 61, 66, 77, 82, 83, 86, 89, 95, 108, 123, 159, 175, 183, 185, 188, 190, 218, 219, 222, 223
Marketing. *See also* Advertisements; Promotion of products
  events, 26, 234
  methods, 9, 18, 156, 216
  specialists, 38, 45, 46, 68, 70, 87, 91, 95, 96, 114, 118, 159, 174, 181, 186
  technologies, 37, 167, 176
Marketplace, 1, 3, 8, 10, 38, 187–190, 234
Massachusetts Institute of Technology (MIT), 2, 30, 38–40, 69, 136, 232
Mauss, Marcel, v, 61, 72, 235
Media Lab, 2, 39, 40, 69, 136, 232
Mediannotation project, 39, 49, 51, 69, 94
Memory, 31, 47, 48, 67, 70, 158, 177, 224
Mermaid mummies exhibit, 152–154, 193, 213
Metaphorical language, use of, 48, 68, 76, 96

Mode: demo, 223, 238n12. *See also* Demo mode
Museums, 32, 142, 152, 213, 215
Mystification, 22, 231. *See also* Illusion; Manipulation: of audiences

Nanotechnology, 129, 163, 164
Narratives, 44, 45, 65, 69, 80, 83, 87, 90, 207. *See also* Storytelling
NEO project, 86–88, 92, 93, 112, 150, 161
NeXT computer, 24, 71, 98, 141, 161, 164
Nongovernmental organizations, 10, 126, 129–131, 163. *See also* Greenpeace
Norms in the practice of demonstrations, 37, 45, 47, 49, 52, 67, 69, 80, 148, 151, 178, 230, 239n3
Nuclear energy, 10, 11, 15, 126–129, 131

Objections from audiences to demonstrations, 52, 112, 232
Opinions
 of experts, 58, 88, 116, 122, 153
 expressed by demonstrators, 4, 5, 7, 88
 of the public, 88, 128, 153, 210, 213
Opposition to projects, 33, 124, 126, 128, 129, 228
Orchestration of demonstrations, 42, 44, 89, 94, 103, 111, 112, 169

Parades, 7, 44
Paris Fair, x, 3, 22, 32, 58, 60, 61, 68, 75, 102, 133, 140, 155, 156, 191
Passion, ix, 70, 74, 205, 236. *See also* Enthusiasm; Euphoria
Passive spectators, 28, 29, 31, 34, 36, 44, 65, 129, 163, 221, 231
Pedagogy, 4, 7, 13, 110, 111, 127, 155, 202, 205, 210, 218. *See also* Didactics; Teaching with demonstrations

Peddlers, 18, 59, 71, 188, 190, 212, 234. *See also* Barkers; Hucksters; Posticheurs
Persuasion, ix, 4, 9, 13, 17, 25, 40, 66, 67, 69, 75, 81, 106, 140, 239n11
Pioneer Zephyr, 25, 98
Playfulness of demonstrations, 2, 47, 69, 194, 198, 204, 210, 211, 224, 225
Play on affects, 68–70, 73–75. *See also* Emotions: play on; Feelings
Plot of demonstrative scenarios, 42, 49, 93, 95
Poetry, 216, 223, 224, 226, 233, 239n11
Police, 42–44, 64, 90–92, 172, 173, 240n5
Posticheurs, 187–189. *See also* Barkers; Hucksters; Peddlers
Powell, Colin, 1, 26, 27, 130
PowerPoint, 1, 8, 15, 28, 37, 47, 48, 66, 67, 101, 114, 117, 130, 172, 180, 216, 217. *See also* Slides
Precautions of demonstrators, 158, 161, 165, 179
Predictions regarding the future of technologies, 29, 71, 162
Predispositions
 of audiences, 134, 140
 of demonstrators, 218
Preparation for demonstrations, ix, 2, 14, 17, 29, 39, 40, 42, 46, 50, 58, 76, 77, 79–83, 87–94, 96, 97, 102, 108, 133, 135, 150, 158, 159, 161, 172, 174, 176, 227, 229, 231, 235, 236. *See also* Rehearsals
Prescriptions for the development of demonstrations, 37, 54, 71, 76, 77, 89, 95, 230. *See also* Advice for the development of demonstrations
Prestidigitation, 25, 51. *See also* Magicians
Processions, x, 3, 15, 29, 30, 198, 199, 234

Products. *See* Comparison of products; Launch of products; Promotion of products; "Vaporware products"
Promises of demonstrators, 24, 81, 99, 170
Promotion of products, 3, 12, 35, 58, 68, 88, 115, 126, 128, 134, 136, 167, 181–184, 194, 203, 206, 209–212, 218, 219, 237n2 (intro). *See also* Advertisements; Marketing
Proof, ix, 4–6, 9–13, 15, 29, 62, 69, 97, 100, 112, 130, 139, 152, 153, 155, 168, 199, 225, 230, 233, 234, 238n7
Prototypes, 26, 30, 31, 38, 81, 86, 90, 94, 106, 107, 109, 122, 124, 161, 170, 218
Prowess, 1, 15, 193, 200–203, 234. *See also* Exploits; Virtuosity
Public experiments, 4, 15, 23, 80, 140, 162

Rehearsals, 21, 79, 80, 82, 83, 95, 150, 156, 160, 172, 229. *See also* Preparation for demonstrations
Relaxation
 of audiences, 25, 71
 of demonstrators, 75, 86, 96, 160, 175
Repertoires, 17, 117, 136, 143, 170, 228
 counterdemonstrative, 171
 demonstrative, 135, 136
 gestural, 77
 of stabilized narratives, 45, 80, 83
Repetition of demonstrations, 48, 51, 54, 74, 79, 83, 118, 195, 234. *See also* Reuse of demonstrations
Reputation, 71, 74, 158, 176
Reuse of demonstrations, 17, 103, 118, 130, 131, 234. *See also* Repetition of demonstrations
Rhetoric, 67, 68, 74, 204, 239n11
Robots, 2, 125, 195, 197
Routines, 67, 72, 86, 173, 185, 191

Sales representatives, x, 3, 18, 38, 59, 72, 156, 172, 177, 178, 185, 186. *See also* Brand ambassadors; Business representatives
Scenarios, 17, 36, 37, 45–54, 58–61, 67, 73, 76, 87, 89, 93, 110, 134, 136, 169, 172, 179, 206, 227, 229, 231. *See also* Scripts
Scenes, 41, 51, 58, 65, 76, 91, 94, 95, 146, 148, 179, 197
Scientists on the March, 3, 42, 63, 64, 69, 75, 90, 91, 102, 113, 114, 200
Scripts, 34, 45, 47, 50, 52, 53, 58–60, 73, 79, 82, 110, 139, 156, 169. *See also* Scenarios
Secrets, 6, 98, 206
Seduction, 71, 72, 160, 169, 204. *See also* Charm
Self-confidence, 28, 71, 158
Sequences. *See also* Duration of performances; Temporality; Timing
 of action, 20, 50, 70, 83, 92, 100, 105, 119, 165, 170, 202
 video, 39, 40, 42, 54, 58, 94, 157, 179, 204–206, 210, 221
Showcases, 3, 24, 25, 95, 174, 181, 193, 194, 234
Silicon Valley, x, 2, 11, 168
Simplicity of demonstrations, 46, 48, 53, 59, 74, 139, 149, 159, 160, 199, 211
Simulations, 16, 19–23, 35, 52, 65, 76, 96, 114, 125, 142
Skepticism, 32, 141, 226. *See also* Doubt
Slickness of demonstrations, 151, 156. *See also* Smoothness of demonstrations
Slides, 4, 48, 67, 81, 118, 127, 217. *See also* PowerPoint
Slogans, 2, 39, 43–45, 48, 54, 59, 63–65, 69, 76, 89, 90, 129, 173, 174, 232
Smiling, as a need for demonstrators, 27, 71, 72, 184. *See also* Laughter

Smoothness of demonstrations, 93, 157, 168. *See also* Slickness of demonstrations
Social class, xi, 21, 74, 142, 190, 198, 199, 235
Social movements, xi, 5, 9, 235
Social networks, 15, 23, 149–151. *See also* Twitter
Solutions, display of, 23, 41, 44, 51, 54, 55, 58, 59, 76, 87–89, 94, 109, 125, 135, 144, 228
Spectators. *See* Attention: of spectators; Dreams of spectators; Imagination of spectators; Passive spectators
Speeches, 25, 26, 43, 44, 47, 61, 64, 66, 68, 70, 72, 73, 75, 86, 185, 222, 227. *See also* Discourse; Talks; Verbal performances
Staging of demonstrations, 1, 10, 17, 24, 25, 28, 36, 41, 43, 45, 48, 68, 71, 80, 91–97, 103, 144, 149, 150, 156, 172, 183, 189, 198, 199, 202, 212–214, 219, 223, 227, 229, 231, 234
Stanford University, 2, 113, 114
Start-ups, 8, 46, 96, 158, 171, 172, 176
Status
 of audiences, 30, 34, 86, 194
 of demonstrations, 10, 26, 152–154, 230
 of demonstrators, 176, 177, 187, 192
Storytelling, 41, 49, 51, 76, 86, 87, 89, 202. *See also* Narratives
Stratification, 33, 175, 187, 192, 198
Substitutions of demonstrations, 11, 61–63, 138, 204, 235
Success stories, 48, 121, 162, 169
Summaries, 52, 62, 65, 101, 108, 109, 116, 117, 120, 121, 123, 204, 205
Supermarkets, 15, 167, 180, 182–184, 191, 234
Surprises, 47–49, 70, 86, 101, 191. *See also* Astonishment of audiences
Sympathy, 71. *See also* Empathy

Talent of demonstrators, 71, 133, 134, 155, 176, 186, 188, 190, 191, 193, 200, 202, 203, 220, 230
Talks, 2, 61, 81, 95, 106, 110, 117, 118, 127, 158, 174, 239n7. *See also* Discourse; Speeches; Verbal performances
Tangibility of demonstrations, 25, 66, 120, 128, 214, 225. *See also* Concreteness of demonstrations
Target audience, 44, 54, 87, 109, 115, 122, 123, 125, 127, 134, 136, 138, 141, 142, 148, 169, 186, 193, 210, 229
Tastings, 60, 73, 181–184, 194
Teaching with demonstrations, 7, 9, 10, 110, 127, 173, 207, 210. *See also* Didactics; Pedagogy
TechDays, 3, 31, 33, 75, 81–86, 96, 100, 101, 112, 133, 144, 145, 148, 149, 151, 156, 159, 161, 174, 175, 194–196, 239n1
"Technical redoings," 20, 21, 35
Technological revolution, 20, 24, 98, 100, 141, 161
Technology evangelist, x, 3, 45, 46, 70, 74, 76, 82, 95, 145, 157, 171, 174, 176, 234, 237n4, 243n2
Telemarketing, 8, 15, 53, 66
Television, 1, 3, 8, 15, 23, 27, 37, 38, 45, 53–55, 58, 69, 95, 96, 127, 128, 183, 212, 234, 240n4
Temporality, 45, 50, 66, 76, 99, 100, 105, 163, 165, 173, 228. *See also* Duration of performances; Sequences; Timing
Testimony, 32, 48, 54, 58, 69, 94, 95, 101, 191, 238n14. *See also* Witnesses
Testing performances on audiences, 9, 79, 85, 87, 89, 109, 110, 143, 146, 171
Test videos, 206, 207, 210, 211

Theater, 1, 3, 4, 23, 49, 93, 95, 96, 198, 199, 212, 217, 220, 237n3
Theatrical performances, 13, 16, 17, 18, 20–22, 24, 28, 35, 47, 76, 93, 95–97, 103, 143, 198, 230, 231
Timing, 41, 43, 48, 61, 90. *See also* Duration of performances; Sequences; Temporality
Tocqueville, Alexis de, 137, 233, 242n6
Tours, 2, 38, 39, 44, 147, 221. *See also* Visits based on demonstrations
Trade shows, 15, 34, 38, 187, 188
Trailers, 48, 50, 168. *See also* Video clips
Training
  of demonstrators, 18, 47, 49, 80, 95, 168, 171–173, 177
  provided by demonstrators, 72, 175, 180, 184–187, 192
Transformation of demonstrations, 101, 117, 118, 131, 212
Transmission of demonstrative skills, 168–170, 172, 173, 190, 191, 230. *See also* Apprenticeship
Tricks, 145, 166, 190, 207
Trust, 58, 60, 71, 160
Tutorials, 1, 110, 204, 207–210
Twitter, 15, 33, 48, 81, 101, 114, 115, 144, 149–151, 156, 159, 161, 194, 210. *See also* Social networks

Unexpected situations, 24, 82, 110, 134, 154, 156, 165, 172, 229
Uniqueness
  of performances, 59, 169, 177
  of what is on display, 54, 58 (*see also* Exceptionality of what is on display)
User conversion. *See* Conversion of users
Utilitarianism, 16, 20, 22, 35, 38, 192, 233
Utopias, 24, 25, 51, 222, 226, 233

Valorization of failures, 82, 157, 161, 165. *See also* Failure of demonstrations

"Vaporware products," 26, 89, 157
Venture capitalists, 38, 46
Verbal performances, 61, 66, 68, 79, 86, 109, 161, 191. *See also* Discourse; Speeches; Talks
Verification of equipment, 158, 172
Versions of technologies, 21, 26, 65, 67, 81, 86, 89, 90, 106, 109, 110, 157, 161, 196
Video clips, 40, 49, 96, 168, 204. *See also* Trailers
Video games, 8, 9, 145, 146, 148, 157, 193, 194, 204–207, 210, 211, 221, 234
Viewpoints, expression of, 5, 6, 11, 62, 63, 68, 113, 143, 151–153, 215, 224. *See also* Vision
Virtual worlds, 66, 123, 195, 204, 225, 233
Virtuosity, 82, 156, 204, 219. *See also* Exploits; Prowess
Visibility, 32, 42, 63, 92–95, 102, 112, 122, 199, 221, 223
Vision. *See also* Viewpoints, expression of
  paradigmatic, 24, 242n10
  utopian, 25
Visits based on demonstrations, 11, 38, 65, 66, 125–128, 143, 167, 180, 240n11, 241n11. *See also* Tours
Visual devices. *See* Charts; Graphics; Visual effects
Visual effects, 41, 47, 49, 64–66, 70, 86, 87, 89, 92, 93, 95, 115, 148, 151. *See also* Charts; Graphics

Witnesses, 20, 29, 34, 35, 44, 107, 118, 141, 162, 238n14. *See also* Testimony

YouTube, ix, 1, 27, 114, 204–211, 220, 234

Infrastructures Series
edited by Geoffrey C. Bowker and Paul N. Edwards

Paul N. Edwards, *A Vast Machine: Computer Models, Climate Data, and the Politics of Global Warming*

Lawrence M. Busch, *Standards: Recipes for Reality*

Lisa Gitelman, ed., *"Raw Data" Is an Oxymoron*

Finn Brunton, *Spam: A Shadow History of the Internet*

Nil Disco and Eda Kranakis, eds., *Cosmopolitan Commons: Sharing Resources and Risks across Borders*

Casper Bruun Jensen and Brit Ross Winthereik, *Monitoring Movements in Development Aid: Recursive Partnerships and Infrastructures*

James Leach and Lee Wilson, eds., *Subversion, Conversion, Development: Cross-Cultural Knowledge Exchange and the Politics of Design*

Olga Kuchinskaya, *The Politics of Invisibility: Public Knowledge about Radiation Health Effects after Chernobyl*

Ashley Carse, *Beyond the Big Ditch: Politics, Ecology, and Infrastructure at the Panama Canal*

Alexander Klose, translated by Charles Marcrum II, *The Container Principle: How a Box Changes the Way We Think*

Eric T. Meyer and Ralph Schroeder, *Knowledge Machines: Digital Transformations of the Sciences and Humanities*

Geoffrey C. Bowker, Stefan Timmermans, Adele E. Clarke, and Ellen Balka, eds., *Boundary Objects and Beyond: Working with Leigh Star*

Clifford Siskin, *System: The Shaping of Modern Knowledge*

Lawrence Busch, *Knowledge for Sale: The Neoliberal Takeover of Higher Education*

Bill Maurer and Lana Swartz, *Paid: Tales of Dongles, Checks, and Other Money Stuff*

Katayoun Shafiee, *Machineries of Oil: An Infrastructural History of BP in Iran*

Megan Finn, *Documenting Aftermath: Information Infrastructures in the Wake of Disasters*

Ann M. Pendleton-Jullian and John Seely Brown, *Design Unbound: Designing for Emergence in a White Water World*, Volume 1: *Designing for Emergence*

Ann M. Pendleton-Jullian and John Seely Brown, *Design Unbound: Designing for Emergence in a White Water World*, Volume 2: *Ecologies of Change*

Jordan Frith, *A Billion Little Pieces: RFID and Infrastructures of Identification*

Morgan G. Ames, *The Charisma Machine: The Life, Death, and Legacy of One Laptop per Child*

Ryan Ellis, *Letters, Power Lines, and Other Dangerous Things: The Politics of Infrastructure Security*

Mario Biagioli and Alexandra Lippman, eds, *Gaming the Metrics: Misconduct and Manipulation in Academic Research*

Malcolm McCullough, *Downtime on the Microgrid: Architecture, Electricity, and Smart City Islands*

Emmanuel Didier, translated by Priya Vari Sen, *America by the Numbers: Quantification, Democracy, and the Birth of National Statistics*

Andrés Luque-Ayala and Simon Marvin, *Urban Operating Systems: Producing the Computational City*

Michael Truscello, *Infrastructural Brutalism: Art and the Necropolitics of Infrastructure*

Christopher R. Henke and Benjamin Sims, *Repairing Infrastructures: The Maintenance of Materiality and Power*

Stefan Höhne, *New York City Subway: The Invention of the Urban Passenger*

Timothy Moss, *Conduits of Berlin: Remaking the City through Infrastructure, 1920–2020*

Blake Atwood, *Underground: The Secret Life of Videocassettes in Iran*

Huub Dijstelbloem, *Borders as Infrastructure: The Technopolitics of Border Control*

Claude Rosental, *The Demonstration Society*

Dylan Mulvin, *Proxies: The Cultural Work of Standing In*